If You're Trying to

Get Better Grades & Higher Test Scores in Science

You've Gotta Have This Book!

Grades 4-6

By Imogene Forte
& Marjorie Frank

Incentive Publications, Inc.
Nashville, Tennessee

Illustrated by Kathleen Bullock
Cover by Geoffrey Brittingham
Edited by Patience Camplair

ISBN 0-86530-646-X

1 2 3 4 5 6 7 8 9 10 08 07 06 05

PRINTED IN THE UNITED STATES OF AMERICA
www.incentivepublications.com

Contents

GET SHARP . . . in Science Concepts & Processes 45

GET SHARP . . . in Space Science 69

GET SHARP . . . in Life Science 133

Get Ready

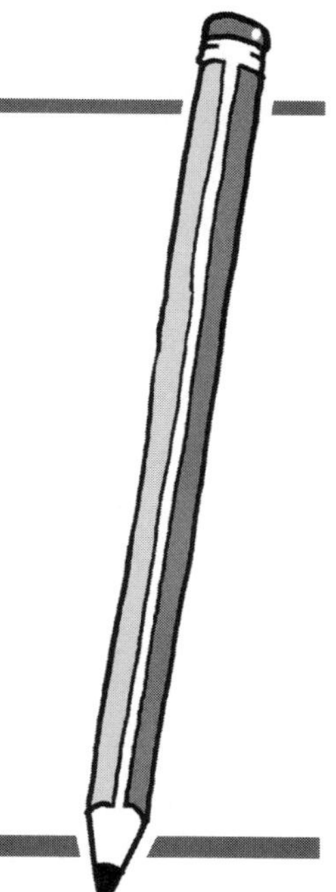

Get ready to get smarter in science. Get ready to be a better student and get the grades you are capable of getting. Get ready to feel better about yourself as a student. Lots of students would like to do better in school. Lots of students CAN do better. But it doesn't happen overnight. The first part of getting ready is: **wanting** to do better—motivating yourself to get moving on this project of showing how smart you really are. The **Get Ready** part of this book will help you do just that: get inspired and motivated. It also gives you some practical ways to organize yourself, your space, your time, and your science homework. And there's more! It gives tips you can use right away to make big improvements in your study habits.

Get Set

Once you have taken a hard look at your goals, organization, and study habits, you can move on to other skills and habits that will help you be more successful at learning. The **Get Set** part of this book starts out with a quick review of some tools you will need for doing science lessons and investigations. Then, it gives an overview of the thinking skills that will help you get the most out of your brain. Top this off with a great review of skills you need for good studying. It's all right here at your fingertips—how to understand and solve science problems, listen well, read carefully, study for tests, and take tests. Take this section seriously, and you will start making improvements immediately in your science performance.

Get Sharp

Now you're ready to mix those good study habits and skills with the science that you need to learn. The **Get Sharp** sections of this book contain all kinds of facts and explanations, processes, definitions, lists, and how-to information. These sections cover the basic areas of science that you study—the nature of science, science history, the big ideas and processes that underlie all branches of science, earth science, space science, life science, human body and health, and physical science. They are loaded with information you need to understand science homework and get it done right. This part of the book is a great reference tool PLUS a *how-to* manual for many science topics and projects. Keep it handy whenever you work any science assignment. And don't overlook the **Get Sharp on Science Terms** section of the book. It is a helpful, complete glossary of all the science terms you will need. Use it to keep those definitions clear!

How to Use This Book

Students

This can be the ultimate science homework helper for you to keep at home. Use the **Get Ready** section to improve your attitude and get motivated to be a good science student. Use the **Get Set** section to sharpen your study skills. Then, have the book nearby at all times when you have science to do at home. Use the **Get Sharp** sections to . . .

. . . strengthen something you've already learned.
. . . get fresh and different examples of something you've studied.
. . . check up on a fact, definition, or detail of science.
. . . get a quick answer to a science question.
. . . clear up something you thought you knew but now aren't sure about.
. . . guide you in a science process (such as how to do a science investigation).
. . . check yourself to see if you've got a fact or process right.
. . . review a topic in preparation for a test.

Teachers

This book can serve multiple purposes in the classroom. Use it as . . .

. . . a reference manual for students to consult during learning activities or assignments.
. . . a reference manual for yourself to consult on particular facts, concepts, or processes.
. . . an instructional handbook for particular science topics.
. . . a complete glossary of science terminology (see **Get Sharp on Science Terms** section).
. . . a remedial tool for anyone needing a review of a particular science process or idea.
. . . a source of advice for parents and students regarding homework habits.
. . . an assessment guide to help you gauge student mastery of science processes or skills.
. . . a source of good resources for building bridges between home and school.

(For starters, send a copy of the letter on page 17 home to each parent. Use any other pages, particularly those in the ***Get Ready*** *and* ***Get Set*** *sections, as send-home pieces.)*

Parents

The **Get Ready** and **Get Set** sections of this book will help you to help your child improve study habits and sharpen study skills. These can serve as positive motivators for the student while taking the burden off of you. Then, you can use the **Get Sharp** sections as a source of knowledge and a process guide for yourself. It's a handbook you can consult to . . .

. . . refresh your memory about a science process, or fact, or get a definition of a science term.
. . . end confusion about kinds of numbers, solving equations, formulas, and many other science questions.
. . . provide useful homework help to your child.
. . . reinforce the good learning your child is doing in school.
. . . gain confidence that your child is doing the homework correctly.

GET READY

Get Motivated

Dear Student,

Nobody can make you a better student. Nobody can even make you WANT to be a better student. But you CAN be. It's a rare kid who doesn't have some ability to learn more, do better with assignments and tests, feel more confident as a student, or get better grades. You CAN DO THIS! You are the one (the only one) that can get yourself motivated.

The first question is this: "WHY would you want be a better student?" If you don't have an answer to this, your chances of improving are not so hot. If you do have answers, but they're like the ones on page 15, your chances of improving still might be pretty slim. Now, we don't mean to tell you that it's a bad idea to get a good report card, or get on the honor roll, or please your parents. But—if you really are going to improve as a student, the reasons need to be about YOU. The goals need to be YOUR goals for your life right now. In fact, if you are having a hard time getting motivated, maybe it is just BECAUSE you're used to hearing a lot of "shoulds" that seem to be about what other people want you to be. Or maybe it's because the goals are so far off in some distant future that it's impossible to stay excited about them.

Why try to be a better student? Consider these reasons:

- to make use of your good mind (Don't miss out on something you could learn to do or understand.)
- to get involved—to change learning into something YOU DO instead of something that someone else is trying to do TO you
- to take charge and get where YOU WANT TO GO (It's YOUR life, after all.)
- to learn all you can for YOURSELF (The more you know, the more you think, and the more you understand—the more possibilities you have for what you can be in your life RIGHT NOW and in the future.)

Follow the "Get Motivated Tips" on the next page as you think about this question. Then, write down a few reasons of your own. These will inspire you to put your brain to work, show how smart you are, and get even smarter.

Sincerely,

Imogene and Marjorie

Why Should I Be A Better Student?

Well, actually, most of these reasons don't motivate me very much at all.

To please my parents?
To please my teachers?
To impress other kids?
To impress my parents' friends?
So people will like me better?
To keep from embarrassing my parents?
To do as well as my brother and sister?
So teachers will treat me better?
To get the money and favors my parents offer for good grades?
To get well-prepared for high school?
To make a lot of money when I finish school?
To get a good report card?
To get into college?

Get Motivated Tips

1. Think about why you want to do better as a student.
2. Think about what you would gain now from doing better.
3. Set some short-term goals *(something you can improve in a few weeks).*
4. Think about what gets in the way of doing your best as a student.
5. Figure out a way to change something that keeps you from improving.

(Use the form on page 16 to record your thoughts and goals.)

What changes could I make in the near future?

Write two short-term goals—things that you could improve in the next month.

1. ______________________________

2. ______________________________

What gets in the way of good grades or good studying for me?

Name the things that most often keep you from doing your best as a student.

1. ______________________________

2. ______________________________

What distraction can I eliminate?

Choose one thing from above that you will try to change or get rid of for the next month.

1. ______________________________

These are my **Get Motivated** goals.

Get Ready Tip #1

Set realistic goals. Choose something you actually believe you can do. Also, you'll have a better chance of success if you set a short time frame for your goal.

Dear Parent:

How can you help your child get motivated to do the work it takes to be a better student? You can't do it for her (or him). But here are some ideas to help students as they find it within themselves to get set to be good students:

- Read the letter to students (page 14). Help your son or daughter think about where she or he wants to go, what reasons make sense to her or him for getting better grades, and what benefits he or she would gain from better performance as a student.
- Help your child make use of the advice on study habits. (See pages 18–26.) Reinforce the ideas, particularly those of keeping up with assignments and turning in work on time.
- Provide your child with a quiet, comfortable, well-lit place that is available consistently for study. Also, provide a place to keep materials, post reminders, display schedules.
- Set family routines and schedules that allow for good blocks of study time, adequate rest, relaxing breaks, and healthy eating. Include some time to get things ready for the next school day and some ways for students to be reminded about upcoming assignments or due dates.
- Keep distractions to a minimum. You may not be able to control the motivations and goals of your child, but you can control the telephone, computer, Internet, and TV. These things actually have on-off switches. Use them. Set rules and schedules.
- Demonstrate that you value learning in your household. Read. Show excitement about learning something new yourself. Share this with your kids.
- Help your child gather resources for studying, projects, papers, and reports. Try to be available to take her or him to the library, and offer help tracking down a variety of sources. Try to provide standard resources in the home (dictionary, thesaurus, computer, encyclopedia, etc.).
- DO help your student with homework. This means helping straighten out confusion about a topic (when you can), getting an assignment clear, discussing a concept or skill, and perhaps working through a few problems along with the student to make sure he or she is doing it right. This kind of help extends and supports the teaching done in the classroom. Remember that the end goal is for the student to learn. Don't be so insistent on the student "doing it himself" that you miss a good teaching or learning opportunity.
- Be alert for problems, and act early. Keep in contact with teachers, and don't be afraid to call on them if you see any signs of slipping, confusion, or disinterest on the part of your child. It is easier to reclaim lost ground if you catch it early.
- Try to keep the focus on the student's taking charge for meeting his or her own goals, rather than on making you happy. This can help get you out of a nagging role and place some of the power in the hands of the student. Both of these will make for a more trusting, less hostile relationship with your child on the subject of schoolwork. Such a relationship will go a long way toward supporting your child's self-motivation to be a better student.

Sincerely,

Imogene and Marjorie

Get Organized

Evan knows a lot about worms. He has read several books and interviewed two scientists. He's bought all the materials for a worm farm. But he's not really able to show what he's learned, because he is so disorganized. Don't repeat Evan's mistakes.

Get Your Space Organized

Get Ready Tip #2

Set up your study space before school starts each year. Make it cozy and friendly—a safe place for getting work done.

Find a good place to study. Choose a place that . . .

. . . is always available to you.
. . . is comfortable and uncluttered.
. . . is quiet and as private as possible.
. . . has good lighting.
. . . is relatively free of distractions.
. . . has a flat surface large enough to spread out materials.
. . . has a place to keep supplies handy. *(See page 19 for suggested supplies.)*
. . . has some wall space or bulletin board space for posting schedules and reminders.

Get Your Stuff Organized

Gather things that you will need for studying or for projects, papers, and other assignments. Keep them organized in one place, so you won't have to waste time looking for them. Here are some suggestions:

Get Ready Tip #3

Have a place to put things you bring home from school. Put your school things in there every time you come in the door—so important stuff doesn't get lost or moved around in the house.

Get set with a place to keep supplies.

(a bookshelf, a file box, a paper tray, a drawer, a plastic dishpan, a plastic bucket, a carton, or plastic crate)

- Keep everything in this place at all times.
- Return things to it after you use them.

Also have:

an assignment notebook
a notebook for every subject
a book bag or pack to carry things back and forth
a schedule for your week (or longer)

Supplies To Have Handy

a good light
a clock or timer
bulletin board or wall
(for schedule & reminders)
pencils, pens, erasable pens
erasers
colored pencils or crayons
markers, highlighters
notebook paper, typing paper
scratch paper
drawing paper
index cards, sticky notes
poster board
folders, report folders
protractor, compass
ruler, measuring tape
calculator
meter stick, yard stick
tape, scissors
glue, glue sticks
paper clips, push pins
stapler, staples

standard references:
science textbook
encyclopedia (set or CD)
homework hotline numbers
homework help websites
good science websites

optional science equipment:
magnets
microscope
telescope
chemistry set

Get Your Time Organized

It might be easy to organize your study space and supplies. But it might not be quite as easy to organize your time. This takes some work. First, you have to understand how you use your time now. Then you will need to figure out a way to make better use of your time. Here is a plan you can follow right away to help you get your time organized.

1. Think about how you use your time now.

For one week, pause at the end of each day, think back over the day, and write down what you did in each block of time for the whole day. Then look at the record you have kept to see how you used your time. Ask yourself these questions:

Did I have any goals for when I would get certain things done?

Did I think ahead about how I would use my time?

How did I decide what to do first?

Did I have a plan or did I just get things done in any order?

Did I get everything done, or did I run out of time?

How much time did I waste?

Get Ready Tip #4

When you plan your week's schedule, don't make it too tight or rigid. Leave room for unexpected events.

I notice that I had only 5 minutes for breakfast every day last week.
Three out of four nights I never got to my science homework.
I started my homework most nights at about 8 o'clock.
I watched an average of three TV shows every night.
I talked on the phone for almost two hours on two nights.

Notice the patterns that helped you get your work done and the patterns that didn't.

2. Make a plan for the next week.

Get a fresh start. Plan your time for next week. Skip the bad habits or poor use of time from last week, and plan to do some things differently. Get a notebook and make a schedule for each day.

Include:

. . . time that will be spent at school

. . . after-school activities

. . . meals

. . . study time

. . . family activities

. . . fun, sports, or recreational activities

. . . social activities or special events

. . . time for rest and sleep

Get Your Assignments Organized

You can't do a very good job of an assignment if you don't have a clue about what it is. You can't possibly do the assignment well if you don't understand the things you are studying. So if you want to get smarter, get clear and organized about assignments. Follow these eight steps:

1. Listen to the assignment.
2. Write it down in an assignment notebook. *(Make sure you write down the due date.)*
3. If you don't understand the assignment—ASK. *(Do not leave the classroom without knowing what it is you are supposed to do.)*
4. If you don't understand the material well enough to do the assignment—TALK to the teacher. (*Tell him or her that you need help getting it clear.*)
5. Take your assignment book home.
6. Write major assignments onto a calendar at home. This way, you can be reminded of projects and other assignments that are coming up.
7. Make a Daily To-Do List *(For each day, write the things that must be done by the end of that day. Make this list a day or two ahead of time.)*
8. Look at your assignment book every day and check your to-do list to make sure you know what needs to be done that day.

Get Ready Tip #5

At all times—keep a copy of class schedules or long-range class assignments at home.

Thurs. TO DO List

Finish Science project
Study for Science test
Read poems for English
Review for spelling quiz
Get soccer bag ready
Return Library books
Review Spanish vocab.

Get Ready Tip #6

Don't count on anyone else to listen to the assignment and get it down right. Get the assignment yourself.

Subject	Assignment	Date Due
English	Check out poetry book	Mon. 11-6
Math	Ch. 2 review problems	Mon. 11-6
Science	Life Science project	Fri. 11-10
English	Grammar quiz	Mon. 11-13
Spanish	Vocab. test	Fri. 11-10
Spelling	quiz, unit 4	Thurs. 11-9
Social St.	Map- So. Am. products	Tues. 11-14

Get Yourself Organized

Okay, so your schedule is on the wall—all neat and clear. Your study space is organized. Your study supplies are organized. You have written down all your assignments, and you've got all your lists made. Great! But do you feel overloaded or stressed? Take some time to think about the behaviors that will help YOU feel as organized as your stuff and your schedule.

Before you leave school . . .

STOP—take a few calm, unrushed minutes to think about what books and supplies you will need at home for studying. ALWAYS take your assignment notebook home.

When you get home . . .

FIRST—Put your school bag in the same spot every day, out of the way of the bustle of your family's activities.

STOP—after relaxing, or after dinner, take a few calm, unrushed minutes to look over your schedule and review what needs to be done. Review your list for the day. Plan your evening study time and set priorities. Don't wait until it is late or you are very tired.

Before you go to bed . . .

STOP—take a few calm, unrushed minutes to look over your assignment notebook and to-do list for tomorrow one more time. Make sure everything is done.

THEN—put everything you need for the next day IN your book bag. Don't wait until the morning. Make sure you have all the right books and notebooks in your bag. Make sure your finished work is all in the bag. Also, pack other stuff (for gym, sports, etc.) at the same time. Put everything in one consistent place, so you don't have to rush around looking for it.

In the morning. . .

STOP—take a few calm, unrushed minutes to review the day one more time.

THEN—eat a good breakfast.

Georgia finished that long science assignment that is due today. She read the chapter and looked up all the terms. She made a diagram of the animal phyla and wrote the characteristics of each group. She was proud of her hard work.

She remembered to take her lunch and gym bag. She took the video game she promised to lend her friend Bob....

. . . but, guess what Georgia forgot to take to school today?

Get Ready Tip #7
It doesn't do much good to get your homework done if you don't turn it in.
!!
Oh, no!

Get Healthy

If you are sick, tired, droopy, angry, nervous, weak, or miserable, it is very hard to be a good student. It is hard to even use or show what you already know. Your physical and mental health is a basic MUST for doing well in school. So, don't ignore your health. Pay attention to how you feel. No one else can do that for you.

Get plenty of rest.

If you're tired, nothing works very well in your life. You can't think, concentrate, pay attention, learn, remember, or study. Try to get eight hours of sleep every night. Get plenty of rest on weekends. If you have a long evening of study ahead, take a short nap after school.

Eat well.

You can't learn or function well on an empty stomach or when your nutrition is poor. Junk food (soda, sweets, chips, snacks) actually will make you more tired. Plus, it crowds the healthy foods out of your diet—the foods your brain needs to think well and your body needs to get through the day with energy. So, eat a balanced diet, with lean meat, whole grains, vegetables, fruit, and dairy products. Drink a lot of water—eight glasses a day is good.

Exercise.

Everything in your body works better when you get a chance to move. Don't let your life become inactive. Do something every day to get exercise—walk, play a sport, play a game, or run. It's a good idea to get some exercise before you sit down to study, too. Exercise helps you relax, unwind, and de-stress. It's good for stimulating your brain.

Relax. Find stress relief.

Your body and your mind need rest. Do something every day to relax. Pay attention to signs of anxiety and stress. Are you nervous, worried, angry, sad, tense? Stress can lower your success in school and interfere with your life. Find a way to relieve the stress. Start with exercise and rest. Also, try these: stretch, take a hot bath or long shower, laugh, listen to calming music, write in a journal. If you are burdened with worries, anger, or problems, talk to someone—a good friend, a teacher or parent, or another trusted adult.

I got plenty of sleep last night. After school, I went to basketball practice and now I'm stretching to relax. After I eat a healthy dinner, I'll be in great shape to study for a science quiz tomorrow.

Get a Grip (on Study Habits)

Here is some good advice for getting set to improve your study habits. Check up on yourself to see how you do with each of these. Then, set goals where you need to improve.

. . . in school:

1. Get to school on time.

When you are late, you get off to a slow start. Sometimes you miss important instructions. Show up on time. Take your book, your notebook, your pencil, and other supplies.

2. Choose your seat wisely.

Sit where you won't be distracted. Avoid people with whom you'll be tempted to chat. Stay away from the back row. Sit where you can see and hear.

3. Pay attention.

Get everything you can out of each lesson. Listen. Stay awake. Your assignments will be easier if you've really been awake and aware during class.

4. Take notes.

List main points. Record examples of problems, solved correctly. If you hear something AND write it, you will be likely to remember it.

5. Ask questions.

It's the teacher's job to see that you understand the material. It's your job to ask if you don't.

6. Use your time in class.

Get as much as possible of the next day's assignment done before you leave school. Use your time during class, between classes, or during study period.

7. Write down assignments.

Do not leave school until you understand the assignment and have it written down clearly.

8. Turn in your homework.

If you turn in every homework assignment, you are a long way toward doing well in a class—even if you struggle with tests.

I'm going to get a head start on my science fair project. It isn't even due until next month!

. . . at home:

1. **Gather your supplies.**
   Before you sit down to study, get all the stuff together that you will need: assignment book, notebook, notes, textbook, study guides, paper, pencils, etc. Think ahead so that you have supplies for long-term projects.

2. **Avoid distractions.**
   Think of all the things that keep you from concentrating. Figure out ways to remove those from your life during study time. Keep your study time uninterrupted.

3. **Turn off the TV. Phone later.**
   No matter how much you insist otherwise, you cannot study well with the TV on. Plan your TV time before or after study time, not during it. The best way to avoid the distraction of the telephone is to study in a room with no phone. Call your friends when your work is done.

4. **Hide the computer games.**
   Stay away from the video game playing stations, computer games, e-mail, and Internet surfing. Plan time for these when studies are done, or before you settle into serious study time.

5. **Plan your time.**
   Think about the time you have to work each night. Make a timeline for yourself. Estimate how much time each task will take, and set some deadlines. This will keep your attention from wandering and keep you focused on the task.

6. **Start early.**
   Start early in the evening. Don't wait until just before bedtime to get underway on any assignment. When it is possible, start the day before or a few days before.

7. **Do the hardest things first.**
   It is a good idea to do the hardest and most important tasks first. This keeps you from avoiding procrastination on the tough assignments. Also, you will be doing the harder stuff when your mind is the most fresh. Study for tests and do hard problems early, when your brain is fresh. Do routine tasks later in the evening.

8. **Break up long assignments.**
   Big projects or test preparations can be overwhelming. Break each long task down into small ones. Then take one small task at a time. This will make the long assignments far less intimidating, and you'll have more successes more often. Never try to do a long assignment all in one sitting.

9. **Take breaks.**
   Plan a break for your body and mind every 30-45 minutes. Get up, walk around, or stretch. Do something active or relaxing.

10. Plan ahead for long-range assignments.

Start early on long-range assignments, big projects, and test preparations. Don't wait until the night before anything is due. You never know what will happen that last day. Get going on long tasks several days before the due date. Make a list of everything that needs to be done for a long-range assignment (including finding information and collecting supplies). Then, start from the due date and work backwards. Make a timeline or schedule that sets a time to complete each of the tasks on the list.

11. Cut out the excuses.

It is perfectly normal to want to avoid doing school work. Remember, however, that excuses take up your energy. In the time you waste convincing yourself or anyone else that you have a good reason for avoiding your studies, you could be getting some of the work done. If you want to be a better student, you will need to dump your own list of excuses.

12. Don't get behind.

Keeping up is good. Many students slip into failure, stress, and hopelessness because they get behind. The best way to avoid all of these is—NOT to get behind. This means DO your assignments on time. If you do get behind because of illness or something else unavoidable, do something about it. Don't get further and further into the pit! Talk to the teacher. Make a plan for catching up. Getting behind is often caused by procrastination. **Don't procrastinate.** The more you put off, the worse you feel, and the harder it is to catch up!

13. Get on top of problems.

Don't let small problems develop into big ones. If you are lost in a class, miss an assignment, don't understand something, or have done poorly on something—act quickly. Talk to the teacher, ask a parent to help, find another student who has the information. Do something to correct the problem before it becomes overwhelming.

14. Ask for help.

You don't have to solve every problem alone or learn everything by yourself. Don't count on someone noticing that you need help. Tell them. Use the adults and services around you to ask for help when you need it. Remember, it is the teacher's job to teach you. Most teachers are happy to help a student who shows interest in getting help.

15. Reward yourself for accomplishments.

If you break your assignments down into manageable tasks, you'll have more successes more often. Congratulate and reward yourself for each task accomplished—by taking a break, getting some popcorn, going for a walk, bragging about what you've done to someone—or any other way you discover. Every accomplishment is worth celebrating!

GET SET

Get Familiar with Science Tools

Formulas, measurement tools, units of measurement—these sound like math tools! They are, but they are also science tools because math is used constantly in science. If you are going to be successful with investigating science questions, you will need to have these firmly planted in your mind.

Know Your Measurement Units

Length

Metric System

1 centimeter (cm) = 10 millimeters (mm)
1 decimeter (dm) = 10 centimeters (cm)
1 meter (m) = 10 decimeters (dm)
1 meter (m) = 100 centimeters (cm)
1 meter (m) = 1000 millimeters (mm)
1 decameter (dkm) = 10 meters (m)
1 hectometer (hm) = 100 meters (m)
1 kilometer (km) = 100 decameters (dkm)
1 kilometer (km) = 1,000 meters (m)

English System (U.S. Customary)

1 foot (ft) = 12 inches (in)
1 yard (yd) = 36 inches (in)
1 yard (yd) = 3 feet (ft)
1 mile (mi) = 5,280 feet (ft)
1 mile (mi) = 1,760 yards (yd)

The total surface area covered by skin on an adult human is about 22 square feet.
The digestive tract of an adult human is about 27 feet long.
Toenails grow approximately 0.1 of a millimeter each day.

Area

Metric System

1 square meter (m^2) = 100 square decimeters (dm^2)
1 square meter (m^2) = 10,000 square centimeters (cm^2)
1 hectare (ha) = 0.01 square kilometer (km^2)
1 hectare (ha) = 10,000 square meters (m^2)
1 square kilometer (km^2) = 1,000,000 square meters (m^2)
1 square kilometer (km^2) = 100 hectares (ha)

English System (U.S. Customary)

1 square foot (ft^2) = 144 square inches (in^2)
1 square yard (yd^2) = 9 square feet (ft^2)
1 square yard (yd^2) = 1,296 square inches (in^2)
1 acre (a) = 4,840 square yards (yd^2)
1 acre (a) = 43,560 square feet (ft^2)
1 square mile (mi^2) = 640 acres (a)

Capacity

Metric System

1 teaspoon (t) = 5 milliliters (mL)

1 tablespoon (T) = 12.5 milliliters (mL)

1 liter (L) = 1,000 milliliters (mL)

1 liter (L) = 1,000 cubic centimeters (cm^3)

1 liter (L) = 1 cubic decimeter (dm^3)

1 liter (L) = 4 metric cups

1 kiloliter (kL) = 1,000 liters (L)

English System (U.S. Customary)

1 tablespoon (T) = 3 teaspoons (t)

1 cup (c) = 16 tablespoons (T)

1 cup (c) = 8 fluid ounces (fl oz)

1 pint (pt) = 2 cups (c)

1 pint (pt) = 16 fluid ounces (fl oz)

1 quart (qt) = 4 cups (c)

1 quart (qt) = 2 pints (pt)

1 quart (qt) = 32 fluid ounces (fl oz)

1 gallon (gal) = 16 cups (c)

1 gallon (gal) = 8 pints (pt)

1 gallon (gal) = 4 quarts (qt)

1 gallon (gal) = 128 fluid ounces (fl oz)

The average dairy cow produces 20 liters of milk daily.
The average American family of four drinks about 280 gallons of milk each year.
In 1998, American dairy cows produced about 21 billion gallons of milk.

The average hot air balloon holds 2,100 cubic meters of hot air.

The volume of Jupiter is over 367 trillion cubic miles.

Volume

Metric System

1 cubic decimeter (dm^3) = 0.001 cubic meter (m^3)

1 cubic decimeter (dm^3) = 1,000 cubic centimeters (cm^3)

1 cubic decimeter (dm^3) = 1 liter (L)

1 cubic meter (m^3) = 1,000,000 cubic centimeters (cm^3)

1 cubic meter (m^3) = 1,000 cubic decimeters (dm^3)

English System (U.S. Customary)

1 cubic foot (ft^3) = 1728 cubic inches (in^3)

1 cubic yard (yd^3) = 27 cubic feet (ft^3)

1 cubic yard (yd^3) = 46,656 cubic inches (in^3)

Weight

Metric System

1 gram (g) = 1,000 milligrams (mg)
1 kilogram (kg) = 1,000 grams (g)
1 metric ton (t) = 1,000 kilograms (kg)

English System (U.S. Customary)

1 pound (lb) = 16 ounces (oz)
1 ton (T) = 2,000 pounds (lb)

Time

Both Systems

1 minute (min) = 60 seconds (sec)
1 hour (hr) = 60 minutes (min)
1 day = 24 hours (hr)
1 week = 7 days
1 year (yr) = 52 weeks
1 year (yr) = 365 or 366 days
1 decade = 10 years
1 century = 100 years

Temperature

The Fahrenheit Scale

freezing point = 32° F
boiling point = 212° F

The Celsius (or Centigrade Scale)

freezing point = 0° C
boiling point = 100° C

The smallest bird in the world is the bee hummingbird. It weighs in at about 1.6 grams.

Bees make 3,300 tons of honey every day.

The sailfish is one of the fastest moving fish in the world. It can swim about 30 meters per second.
It is about 30 million degrees Fahrenheit in the center of the Sun.

The temperature of dry ice is -109.3° F.

Know Your Measurement Equivalents

The sound of a growling bear travels to your ears at about 1,100 feet per second.

That's 335 meters per second!

The tiny desert rat can leap 15 feet. (That's 4.5 meters!)

From English to Metric

English Customary Unit	*Approximate Metric Equivalent*
1 inch	2.45 centimeters
1 foot	30.48 centimeters
1 yard	0.9144 meters
1 mile	1.609 kilometers
1 acre	4,047 square meters
1 ounce	28.3495 grams
1 pound	453.59 grams
1 ton	907.18 kilograms
1 pint	0.4732 liters
1 quart	0.9465 liters
1 gallon	3.785 liters
1 bushel	35.2390 liters

From Metric to English

Metric Unit	*Approximate English Equivalent*
1 millimeter	0.04 inch
1 centimeter	0.39 inch
1 meter	39.37 inches
1 kilometer	3,281 feet or .62 miles
1 gram	0.0353 ounce
1 hectogram *(100 grams)*	3.53 ounces
1 kilogram	2.2 pounds
1 metric ton	2,046.6 pounds or 1.1 tons
1 liter	1.06 quarts

Temperature Conversions

to change Fahrenheit to Celsius: subtract 32, then multiply by $\frac{5}{9}$

to change Celsius to Fahrenheit: multiply by $\frac{9}{5}$, then add 32

Know Your Formulas

Get Set Tip #1

Memorize these letters and symbols, so you will always know what they mean in formulas.

Perimeter

$P = s + s + s$	Perimeter of a triangle
$P = 2(h + w)$	Perimeter of a rectangle
$P =$ sum of sides	Perimeter of irregular polygons
$C = 2\pi r$	Perimeter or circumference of a circle
$C = \pi d$	Perimeter or circumference of a circle

Area

$A = \pi r^2$	Area of a circle
$A = s^2$	Area of a square
$A = bh$	Area of a parallelogram
$A = \frac{1}{2}bh$	Area of a triangle
$A = \frac{1}{2}(b_1 + b_2)h$	Area of a trapezoid

Letters & Symbols

- h = height
- w = width
- b = base
- B = area of base
- s = side
- π = pi (3.14)
- r = radius
- d = diameter

Volume or Capacity

$V = Bh$	Volume of a rectangular or triangular prism
$V = \frac{1}{3}Bh$	Volume of a pyramid
$V = s^3$	Volume of a cube
$V = \pi r^2 h$	Volume of a cylinder
$V = \frac{1}{3}\pi r^2 h$	Volume of a cone
$V = \frac{4}{3}\pi r^3$	Volume of a sphere

A tennis court is 36 feet wide and 78 feet long.

A tortoise crawling around the perimeter would have to travel 228 feet.

His speed is 0.25 mph. If he travels without stopping, he could get around the court in about 10 minutes.

More Formulas

Work =
force times distance
$W = F \times d$

Speed (velocity) **=**
distance divided by time
$V = \frac{d}{t}$

Force =
mass times acceleration
$F = m \times a$

Power =
work divided by time
$P = \frac{W}{t}$

Mechanical Advantage =
resistance force divided by effort force
$MA = \frac{F_r}{F_e}$

On a safari, Aunt Lucy is chased by an elephant. Can she outrun the elephant?

A hairdryer uses 1500 watts of power. An eletric toothbrush uses 7 watts.

An African elephant can travel at a velocity of 40 kilometers per hour. This is faster than the fastest human.

How much power will the generator have to produce for Aunt Lucy to dry her hair and brush her teeth in the desert each morning?

Electrical Power =
voltage times current
$P = V \times I$

Electrical Energy =
the power delivered times the length of time it is used.
$E = P \times t$

Electrical Current =
voltage divided by resistance
$I = \frac{V}{R}$

Get Tuned-Up on Thinking Skills

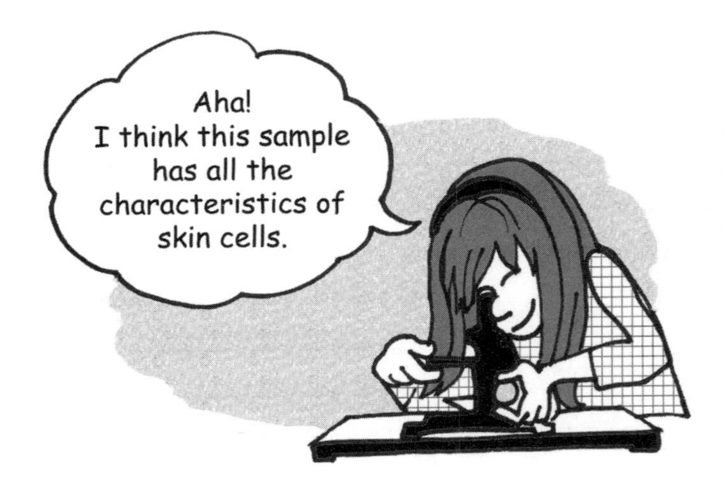

Aha!
I think this sample
has all the
characteristics of
skin cells.

Your brain is capable of an amazing variety of accomplishments! There are different levels and kinds of thinking that your brain can do—all of them necessary to get you set for good learning. Here are some of the thinking skills that are frequently used in doing science tasks and the math tasks involved in science. Use this information to exercise your brain. Then put it to good use as you learn science concepts, use science processes, and investigate science problems.

Recall — To **recall** is to know and remember specific facts, names, processes, categories, ideas, generalizations, theories, or information.

Examples: *Recall helps you remember such things as: the characteristics of the three forms of matter, the names and locations of the planets, the groups and subgroups in the system of life classification, Newton's laws of motion, the names of the bones in the human body, or how to conduct a scientific investigation.*

Classify — To **classify** is to put things into categories. When you classify ideas, numbers, topics, or things, you must choose categories that fit the purpose and clearly define each category.

Example: *octopus, snail, clam, oyster, conch, squid*

There are many different ways to classify these items. They are all organisms. They are all animals. They are also all sea creatures, mollusks, and invertebrates.

Generalize — **Generalizing** is making a broad statement about a topic based on observations or facts. A generalization should be based on plenty of evidence (facts, observations, and examples). Just one exception can prove a generalization false.

Example of Safe Generalization:
Climates at high latitudes are likely to be cooler than those at low latitudes.

faulty generalization (false) — A faulty generalization is invalid because there are exceptions.
Example: *Average temperatures are lower at latitudes of 45° than at latitudes of 25°.*

broad generalization (false) — A broad generalization suggests something is *always* or *never* true about *all* or *none* of the members of a group. Most broad generalizations are untrue.
Example: *The climate is hot and dry at all locations between 0° and 10° latitude.*

Elaborate — To **elaborate** is to provide details about a situation (to explain, compare, or give examples). When you elaborate, you might use words or phrases such as these: *so, because, however, but, an example of this is, on the other hand, as a result, in addition, moreover, for instance, such as, if you recall, furthermore, another reason is.*

Example: *Molecules move farther apart when a substance is heated. For instance, the molecules in ice cubes will move apart as the ice sits in the sun. This will cause the ice to turn into a liquid.*

Get Set Tip #2

Thinking skills are rarely used in isolation from one another. For example: in order to classify things, you must be able to compare and contrast.

Predict — To **predict** is to make a statement about what will happen. Predictions are based on some previous knowledge, experience, or understanding.

Example: *The trees on the hill behind Tom's home burned last summer. All the underbrush also burned. This left bare, dusty ground. Tom predicts that there will be erosion and mudslides when the rains come this fall.*

Infer — To **infer** is to make a logical guess based on information.

Example: *The plants out in Ann's yard were dead on a cold morning. The plants close to the house looked healthy. She inferred that frost had settled on the plants in the yard.*

Recognize Cause and Effect — When one event occurs as the result of another event, there is a **cause-effect relationship** between the two. Recognizing causes and effects takes skill. When reading a science equation or problem, pay careful attention to words that give clues to cause and effect (*the reason was, because, as a result, consequently, so*).

Example: *Lulu stuck her hand into the lion's cage.* (cause)

As a result, she has a large scratch on her hand. (effect)

Hypothesize — To **hypothesize** is to make an educated guess about a cause or effect. A hypothesis is based on examples that support it but do not prove it. A hypothesis is something that can be—and should be—tested.

Example: *A steel ball will fall a distance of 20 feet faster than a feather.*

Extend — To **extend** is to connect ideas or things together, or to relate one thing to something different, or to apply one idea or understanding to another situation.

Example: *You learn that a crab is a crustacean because it has a segmented body with two main regions, a hard exoskeleton, two pairs of antennae, and claw-like legs at the anterior end. You decide that a lobster (which has the same characteristics) is also a crustacean.*

Compare & Contrast — When you **compare** things, you describe similarities. When you **contrast** things, you describe the differences.

Example: ***Compare*** *a diamond and carbon dioxide. Both are substances made of atoms. Both contain carbon.*

Contrast *them. A diamond is a solid; carbon dioxide is a gas. A diamond is hard; carbon dioxide is soft. Oxygen is present in carbon dioxide but not in a diamond.*

Draw Conclusions — A **conclusion** is a general statement that someone makes after analyzing examples and details. A conclusion generally involves an explanation someone has developed through reasoning.

Example: *Maxie bangs on bottles with various amounts of liquid in them. She notices that there are different pitches from the different bottles. She notices that the bottle with 1 inch of liquid gives a lower pitch than the bottle with about 5 inches of liquid. She draws these conclusions:*

1) The amount of air in a bottle affects the sound made when banging on the bottle.

2) The greater the amount of air in the bottle, the lower the pitch that results from tapping.

Analyze — To **analyze,** you must break something down into parts and determine how the parts are related to each other and how they are related to the whole.

Examples: *You must analyze to . . .*

. . . identify the different organs in the digestive system and describe their functions.

. . . explain the difference between characteristics of an insect and an arachnid.

. . . discuss the functions of different organisms in an ecosystem.

Synthesize — To **synthesize,** you must combine ideas or elements to create a whole.

Examples: *You must synthesize to . . .*

. . . understand how muscles, tendons, ligaments, and bones work together to move the body.

. . . understand how salt and warm water mix together to form a solution.

. . . create a graph to show the results of an experiment.

Ouch! Ouch! Ouch!
I've just been hit by a meteor!

Think Logically (or Reason) – When you think **logically**, you take a statement or situation apart and examine the relationships of parts to one another. You reason **inductively** *(start from a general principle and make inferences about the details)* or **deductively** *(start from a group of details and draw a broad conclusion or make a generalization).*

Example: *A meteoroid is a small asteroid that orbits the sun.*
A meteoroid becomes a meteor when it enters Earth's atmosphere.
If a meteor collides with Earth's surface, it is called a meteorite.

No!
Chester has deduced incorrectly. The thing that hit his foot **must** have been a meteorite.

Evaluate — To **evaluate** is to make a judgment about something. Evaluations should be based on evidence. Evaluations include opinions, but these opinions should be supported or explained by examples, experiences, observations, and other forms of evidence.

Examples: *When you evaluate an argument, an explanation, a decision, a prediction, an inference, a conclusion, or a generalization, ask questions like these:*

Are the conclusions reached based on good examples and facts?
Is there good evidence for the generalization or inference?
Is this believable?
Does the explanation make sense?
Are the sources used to make the decisions reliable?
Is the argument effective?
Is it realistic?

Get Serious About Study Skills

Better Listening

Get Set Tip #3
Stop talking! (You can't listen while you talk!)

Keep your ears wide open! You can increase your understanding of science concepts and processes if you listen well. Here are some tips for smart listening. They can help you get involved with the information instead of letting it just buzz by your ears.

1. Realize that the information is important.

Here's what you can get when you listen to someone who is talking to you about science:

. . . details about how a science process works
. . . help answering science questions
. . . examples of problems solved correctly or investigations done correctly
. . . hazards or difficulties you might face when carrying out a science process
. . . new, exciting information about how something in the world works
. . . meanings of terms used in science questions or assignments
. . . directions for certain assignments

2. Be aware of the obstacles to good listening.

Know ahead of time that these will interfere with your ability to listen well. Try to avoid them, alter them, or manage them, so they don't get in the way.

. . . tiredness
. . . surrounding noise
. . . uncomfortable setting
. . . personal thoughts or worries
. . . wandering attention
. . . too many things to hear at once
. . . missing the beginning or ending
. . . talking

3. Make a commitment to improve.

You can't always control all obstacles (such as the comfort of the setting or the quality of the speaker's presentation), but there are things you can control. Put these to work to gain more from your listening.

- Get enough rest.
- Do your best to be comfortable while you listen.
- Cut out distractions. Keep your mind focused on what is being said.
- Look directly at the speaker.
- Take notes. Write down main points the speaker shows or emphasizes.
- As the speaker talks, think of examples or relate the information to your life.
- Pay special attention to opening and closing remarks and anything that is repeated.

Careful Reading

There is plenty of reading in science. Textbooks and other materials explain science concepts and processes. Many science problems are more than just facts or numbers—they include ideas that you need to interpret. All science questions or assignments include some sort of instructions to follow. So, to succeed in science, you need to make good use of reading skills.

Before you read a science assignment or problem, have a clear idea of the purpose for reading.

Are you reading to find directions for the assignment?

Are you reading to learn how to do a science process?

Are you reading to solve a problem?

Are you reading to find a particular fact?

In all these cases above, you need to read closely and carefully. To learn or review a process, or to solve a problem, you will need to read the information more than once.

Read the information about comets. Find out what materials make up each part of the comet. Underline any sentence that helps to answer the question at the end of the paragraph.

A ***comet*** is a large clump of ice, dust, and frozen gases that orbits the Sun. As the comet nears the Sun, some of the ice vaporizes, creating a spectacular tail. The nucleus of the comet is mostly solid, containing ice, dust, solid particles, and gas. The coma is a thick cloud of water and gases that surround the nucleus. This can be thousands of times wider than the nucleus. The tail is made of dust and gas. It can trail behind the comet for millions of miles. The tail is visible when sunlight shines through it.

What substance can be found in all parts of the comet?

If you read the problem carefully, you will . . .

. . . find clear directions for the assignment.

Read the information about comets.

Find what materials make up each part.

Underline any sentences that help answer the question.

. . . identify the question that must be answered.

What substance can be found in all parts of the comet?

. . . find the information needed to help you decide how to answer the question.

The nucleus contains gas.

The coma contains gases.

The tail contains gas.

How to Prepare for a Test

Get SetTip #4
Don't wait until the last minute to study for a test!

Good test preparation does **not** begin the night before the test.

The time to get ready for a test starts long before this night.

Here are some tips to help you get ready—weeks before the test and right up to test time.

1. Start your test preparation at the beginning of the year—or at least as soon as the material is first taught in the class.

The purpose of a test is to give a picture of what you are learning. That learning doesn't start the night before the test. It starts when you start attending the class. Think of test preparation this way, and you'll be less overwhelmed or nervous about an upcoming test.

You'll be much better prepared for a test *(even one that is several days or weeks away)* if you . . .

. . . *pay attention in class*
. . . *work out sample problems*
. . . *keep your notes and class handouts organized*
. . . *read all your assignments*
. . . *do your homework regularly*
. . . *make up any work you miss when you're absent*
. . . *ask questions in class about anything you don't understand*
. . . *review notes and handouts regularly*

Study Schedule
I have a full study schedule.

2. Once you know the date of the test, make a study plan.

- Look over your schedule and plan time to start organizing and reviewing material.
- Allow plenty of time to go through all the material.
- Your brain will retain more if you review it a few times and spread the studying out over several days.

3. Get all the information you can about the test.

- Make note of everything the teacher says about the test.
- Get clear about what material will be covered.
- If you can, find out about the format of the test.
- Make sure you get all study guides the teacher distributes.
- Make sure you listen well to any in-class reviews.

4. Use your study time effectively.

Dos & Don'ts

Do gather and organize all your notes and handouts.

Do review your text. Pay attention to bold words, bold statements, and examples, operations, or problems.

Do identify the kinds of problems in the section being tested; practice solving a few of each kind.

Do review the questions at the end of text sections; practice answering them.

Do review the study guides provided by the teacher.

Do review any previous quizzes on the same material.

Do make study guides and aids for yourself.

Do make sets of cards with key vocabulary words, terms and definitions, main concepts, and types of problems.

Do ask someone (reliable) to quiz you on the main points and terms.

Don't spend your study time blankly staring at your notebook or mindlessly leafing through your textbook.

Don't study with someone else unless doing so with that person actually helps you learn material better.

Don't study in blocks of time so long that you get tired, bored, or distracted.

5. Get yourself and your supplies ready.

Do these things the night before the test (not too late):

Gather all the supplies you need for taking the test (good pencils with erasers, erasable pens, scratch paper, calculator with batteries).

Put these supplies in your school bag.

Gather your study guides, notes, and text into your school bag.

Get a good night of rest.

In the morning:

Eat a healthy breakfast.

Look over your study guides and note card reminders.

Relax, and be confident that your preparation will pay off.

Are these the right substances to be mixing together? I should have reviewed my notes.

How to Take a Test

Before the test begins . . .

- Have supplies ready: take sharpened pencils, scratch paper, calculator, eraser.
- Try to get a little exercise before class to help you relax.
- Go to the bathroom and get a drink.
- Arrive at the class on time (or a bit early).
- Get settled into your seat; get your supplies out.
- If there's time, you might glance over your study guides while you wait.
- To relax, take some deep breaths; exhale slowly.

When you get the test . . .

- Put your name on all pages.
- Before you write anything, scan over the test to see how long it is, what kinds of questions it has, and generally what it includes.
- Think about your time and quickly plan how much time you can spend on each section.
- Read each set of directions twice. Circle key words in the directions.
- Answer all the short-answer questions. Do not leave any blanks.
- If you are not sure of an answer, make a smart guess.
- Don't change an answer unless you are absolutely sure it is wrong.

I'm not sure of the answer. So I will make a smart guess. I'll put an X by this problem so I'll remember to come back to it later.

Get Set Tip #5

The first answer that comes to your mind is correct more often than not. So stick with it unless you are positive about another answer.

Let's see . . . I am still not sure about this question, so I will stay with my first impression.

More Test-Taking Tips

Tips for Solving Science Word Problems Involving Math

Get Set Tip #6
On all questions, on all tests—always read the directions through twice.

A **word problem** uses words to describe a problem or question which needs a solution.

- First, read the problem twice.
- Identify the question to be answered. Underline it.
- Circle key facts needed to solve the problem.
- Circle clue words that point to the correct operation.
- Choose a strategy for solving the problem.
- Write down a problem or equation that could solve it.
- Draw diagrams, charts, or pictures if you need them.
- Solve the problem.
- Go back and read the problem again. Ask yourself:

  Did I answer the question that the problem required?

  Does the answer make sense?
- Check your answer using another method or strategy, if you have time.

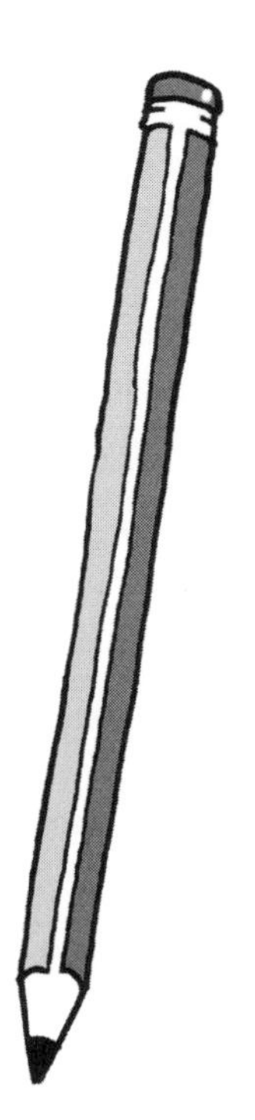

Tips for Solving Number Sentences or Equations

A **number sentence or equation** is a problem made up of numbers, usually with a missing number for you to find. In an equation, a letter often represents the missing number.

- Read the number sentence twice.
- Identify the missing number (or element) that you need to find.
- Simplify the sentence or equation by combining elements that are the same, or by doing easy computations.
- Identify the operation needed to solve the problem.
- Solve the problem.
- Write your answer into the number sentence or equation.
- Read the problem again to make sure it is correct with your answer inserted.

Even More Test-Taking Tips

Tips for Answering Multiple Choice Questions

Multiple choice questions give you several answers from which to choose.

- Read the question twice.
- Before you look at the choices, close your eyes and answer the question. Then look for that answer.
- Read all the choices before you circle one.
- If you are not absolutely sure, cross out answers that are obviously incorrect.
- Choose the answer that is most complete or most accurate.
- If you're not absolutely sure, choose an answer that has not been ruled out.
- Do not change an answer unless you are absolutely sure of the correct answer.

Tips for Answering Matching Questions

Matching questions ask you to recognize facts or definitions in one column that match facts, definitions, answers, or descriptions in a second column.

- Read through both columns to get familiar with the choices.
- Do the easy matches first.
- Cross off answers as you use them.
- Match the leftover items last.
- If you don't know the answer, make a smart guess.

Tips for Answering Fill-in-the Blank Questions

Fill-in-the-blank questions ask you to write a word that completes the sentence.

- Read through each question. Answer it the best you can.
- If you don't know an answer, **X** the question and go on to do the ones you know.
- Go back to the **X'**d questions. If you don't know the exact answer, write a similar word or definition—come as close as you can.
- If you have no idea of the answer, make a smart guess.

Tips for Answering True-False Questions

True-False questions ask you to tell whether a statement is true or false.

- Watch for words like ***most, some,*** and ***often.*** Usually statements with these words are TRUE.
- Watch for words like ***all, always, only, none, nobody,*** and ***never***. Usually statements with these words are FALSE.
- If any part of a statement is false, then the item is FALSE.

GET SHARP

in

SCIENCE CONCEPTS & PROCESSES

The Science Times

The Newspaper of Science History & Ideas

Science is a way of learning about the natural world.

Scientists study many different substances and topics, but they all explore the way the world works. This way of learning called *science* has some special characteristics to know.

Science Has Existed Throughout History and Across Cultures

- Scientific discoveries have been taking place for as long as human life has existed.
- Men and women of many different cultures, countries, ages, and ethnic backgrounds have made important contributions to science.
- Scientists have a wide range of interests, skills, professions, and abilities.
- Scientists work in many different places and ways, often in cooperation with other scientists.
- You don't have to be a professional scientist to think scientifically or to investigate a question with a scientific approach.

The Wheel is Over 5500 Years Old

Almost six thousand years ago, a simple wooden wheel changed transportation forever. Now wheels made of many different substances travel at speeds up to hundreds of miles per hour.

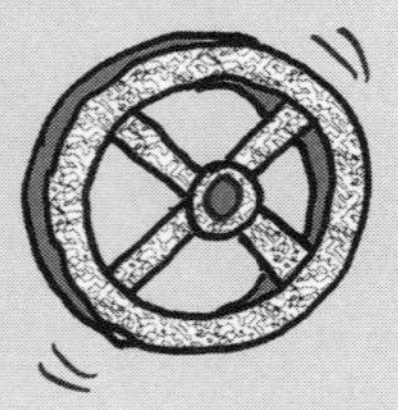

Science Is a Human Endeavor

- Science is done by humans. This means it is influenced by human motivations, opinions, interpretations, beliefs, values, and biases.
- Good scientific research requires many human qualities such as curiosity, creativity, honesty, and judgment.

News Flash!
Congress debates the ethics of stem-cell research.

Science and Society Are Connected

- Science affects the daily lives of humans in thousands of ways.
- Scientific advances have long-lasting effects on people, society, and the planet.
- Scientific research and the uses of science are affected by the needs, values, and politics of a society.
- Science in itself is not good or bad, but uses of science may have benefits or harmful consequences (or both).

Secretary's Invention Solves Chronic Problem

In the 1950s, a single working mother had a great idea. Like other secretaries, Bette Nesbith Graham was bothered by the mess made when she erased mistakes in typing. So she created a brush-on correction fluid, now known as ***Liquid Paper***.

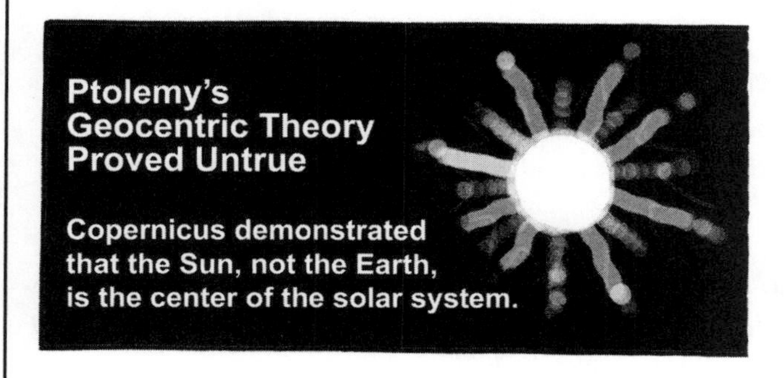

Ptolemy's
Geocentric Theory
Proved Untrue
Copernicus demonstrated

Scientific Knowledge Is Constantly Changing

- All scientific ideas, laws, theories, and knowledge are subject to change and growth.
- Scientific knowledge is always building on earlier knowledge.
- Scientific knowledge changes as new evidence is formed. Old theories are revised; new theories develop.

NEW ELEMENT DISCOVERED

The Periodic Table will need to be revised again, after the production of element 110 in 1994. In 2003, the element was named Darmstadtium.

Do Cell Phones Cause Cancer?

Concerned citizens and consumer advocates wonder about the safety of cellular phones. Scientific research has not yet fully answered the question.

Breaking News!
New bacteria is resistant
to all known antibiotics.

Science Involves a Particular Way of Knowing Things

- Scientists and other curious persons investigate to find out how the world works.
- Scientific investigation involves observation, imagination, and logical thinking.
- Scientists use careful methods to collect facts and observations. Then they use these to build theories that explain how or why things happen.
- **Evidence** is very important in science.
- A theory is accepted after someone else tests it. To become a part of scientific knowledge, the same results must be found from repeated tests.
- Scientists publish reports on their investigations so other scientists can review them. (Review by an outside party is called *peer review*.)
- In science, any result or theory is open to close examination and more questions.
- Many scientific conclusions and explanations raise more questions.

Science Has Limits!

Science can answer or solve only some questions or problems, not all of them. There are many human problems that science cannot solve. Scientific advances can actually create some problems.

SCIENTISTS CLONE SHEEP!

Science & Technology

Science is around us everywhere, every day—mostly because of the technology that science has made possible. Modern technology affects just about every part of our lives—the way we travel, move, eat, play, communicate, learn, and work.

- Scientific experiments and discoveries lead to the invention and creation of technological tools.
  For example: The discovery of electricity made many machines and instruments possible (such as the electromagnet, the vacuum cleaner, the light bulb).

- On the other hand, technological tools or instruments are needed to help scientists make discoveries and answer questions.
  For example: The telescope made it possible for astronomers to discover planets. Powerful microscopes made it possible for scientists to learn about the existence of atomic particles.

- Science and technology also play a part in many issues outside our homes—local, national, and global.

- The choices that people (or societies) make about how to use science have long-lasting effects on individuals, natural resources, and ecosystems. *(Think about choices such as whether to permit widespread cutting of trees in the rainforest, how much control a government should have over the Internet, or how to balance all the needs for water in a county or region.)*

- Most instruments of technology can provide **benefits** for human life. At the same time, these technologies have **costs** or **negative consequences**. *(Think about the benefits and costs of some technologies such as fast racing skis, sound amplifiers, DVD players, jet engines, microwave ovens, video games, computers, gunpowder, roller coasters, or chain saws.)*

- Humans must constantly make judgments and ethical decisions about how to put science to use.

Here are just a few examples of ways a technology can spell good news or bad news for humans, or for the planet Earth.

good news ▸ bad news

Credit Cards

Good News:

- convenience
- don't have to carry cash
- fuels the economy
- buy now, pay later

Bad News:

- people get deeply into debt
- large interest payments
- people buy more than they need
- buy now, pay later

Germs Produced in Lab

Good News:

- scientists can create vaccines to protect against disease

Bad News:

- labs can create germs for biological warfare

City Bus System

Good News:

- convenient way to get around
- reduces pollution from cars
- cheaper than owning a car
- safer than walking in some areas

Bad News:

- noise pollution
- exhaust pollution
- danger to pedestrians
- keeps people from healthy walking

Dams on Mississippi

Good News:

- keep the river from flooding and damaging property

Bad News:

- interfere with the natural processes of river flow and river deposits
- make cities and land vulnerable to flooding from a tide surge caused by a hurricane

Automatic Saws

Good News:

- far fewer mill employees needed
- greater level of safety
- owners save on wage costs
- saws are fast (increased production)

Bad News:

- job loss for workers
- saws cost money, may increase lumber prices
- use up natural resources for power

Theories & Laws

Some Key Theories

Big Bang Theory – The universe formed as the result of a giant, violent explosion.

Cell Theory – The cell is the basic structural and functional unit of all plants and animals.

Continental Drift Theory – The continents were once a single land mass, but have moved from their original locations.

Electromagnetic Theory – Electric and magnetic fields act together to produce electromagnetic waves of radiant energy.

Germ Theory – Infectious diseases are caused by microorganisms.

Heliocentric Theory – Earth and the other planets revolve around the Sun.

Plate Tectonics Theory – The Earth has an outer shell of rigid plates that move about on a layer of hot, flowing rock.

Quark Theory – The nuclear material of atoms is made up of subatomic particles.

Theory of Evolution – All species of plant and animal life developed gradually from a small number of common ancestors.

Theory of Relativity – Observations of time and space are relative to the observer.

Theory of Superconductivity – The electrical resistance of a substance disappears at very low temperatures.

What's The Difference?

Some Key Scientific Laws & Principles

Archimedes' Principle – The loss of weight of an object in water is equal to the weight of the displaced water.

Beer's Law – No substance is perfectly transparent, but some of the light passing through the substance is always absorbed.

Bernoulli's Principle – The pressure of a fluid increases as its velocity decreases, and decreases as its velocity increases.

Boyle's Law – Decreasing the volume of a gas will increase the pressure that the gas exerts (if the temperature does not change).

Charles' Law – The volume of a gas increases as its temperature increases (if the pressure does not change).

Law of Conservation of Matter – Matter is neither created nor destroyed in a chemical change. It is only rearranged.

Law of Hydrostatics – The pressure caused by the weight of a column of fluid is determined by the height of the column.

Law of Universal Gravitation – A gravitational force is present between any two objects. The size of the force depends on the masses of the two objects and the distance between the two objects.

Mendel's Laws – Units called genes, which occur in pairs, determine heredity characteristics in living organisms.

Newton's First Law of Motion **(Law of Inertia)** – A mass moving at a constant velocity tends to continue moving at that velocity unless an outside force stops the movement or causes a change in direction.

Newton's Second Law of Motion **(Law of Action)** – The acceleration of an object depends upon its mass and the applied force.

Newton's Third Law of Motion **(Law of Reaction)** – For every action there is an equal and opposite reaction.

Pascal's Law – Pressure that is applied to a fluid enclosed in a container is transmitted with equal force throughout the container.

Principle of Uniformitarianism – The processes that act on Earth's surface today are the same as the processes that have acted upon Earth's surface in the past.

Different Kinds of Science

Branches of Science

Scientific study is generally divided into major groups, called branches. Many smaller categories of study (fields) exist within each branch.

The Physical Sciences focus on the structure, properties, and behaviors of matter (non-living).

The Life Sciences focus on the structure and function of living organisms and their interaction with the physical environment.

The Earth & Space Sciences focus on the composition, history, forces, and movements of Earth, the solar system, and the universe.

The Social Sciences focus on the people, groups, and institutions of human society, examining the relationships between individuals and the social groups to which they belong.

Mathematics & Logic focus on reasoning and processes that need numbers (measurements, calculations, predictions).

Fields of Science

What is studied within each of these fields?

aerodynamics — mechanics of motion between air and a solid in air

algebra — solving equations

anatomy — parts of organisms and their relationships to each other

anthropology — human cultures

archeology — past human cultures and the items made and used by the cultures

arithmetic — numbers and methods for calculating with numbers

astronomy — objects in space and the history and structure of the universe

astrophysics — physical processes in the workings of objects in space

atomic physics — structure and properties of atoms

bacteriology — bacteria

biochemistry — chemical makeup of living things and life processes

biology — living things

biophysics — physical processes in the functions of living things

botany — plants

chemistry — the composition, structure, and changes of substances

climatology — Earth's climates (weather patterns and trends)

cryogenics — behavior of matter at very low temperatures

cytology — structures of cells

ecology — relationships of living things to each other and to their environment

economics — systems by which people produce, distribute, and use goods and services

electronics — application of scientific technology of electricity

entomology — insects and insect control

genetics — heredity and genes

geochronology — age and history of the Earth and its parts

geography — physical structures of Earth's surface and their relationship to human life

geology — the history, composition, and structure of the Earth

geometry — relationship of points, lines, surfaces, figures, and solids in space

herpetology — reptiles

histology — tissues

ichthyology — fish

immunology — immune system

marine biology — all life in the sea

medicine — study, prevention, and treatment of disease, injury, and sickness

meteorology — Earth's atmosphere and weather

microbiology — organisms that can be seen only with the aid of a microscope

mineralogy — minerals

nuclear physics — structure, composition, and behavior of the nuclei of atoms

oceanography — physical properties, processes, and inhabitants of the ocean

organic chemistry — compounds containing the element carbon

ornithology — birds

paleontology — forms of life from prehistoric times

pathology — diseases

petrology — rocks

physics — matter and energy

political science — politics, power, and governments

probability — the likelihood that an event will occur

psychology — human behavior

physiology — functions of living things and their parts

radiology — X-rays and other forms of radiant energy

seismology — earthquakes

sociology — interrelationships of people in human societies and communities

statistics — analysis of large amounts of numerical data

taxonomy — classification of living things

thermodynamics — heat as produced by the motion of molecules

zoology — animals

I guess we're the reason they call it a zoo.

Some Dazzling Events

These are just a few of the remarkable, fascinating, life-changing discoveries, inventions, and events that mark the history of science. You can do further research to learn more about any of these.

B.C.

8000 domestication of animals

4000 first use of irrigation

4000 invention of the iron-smelting process

3000 first mapping of the stars

2600 invention of geometry

700 creation of first calendar

200 invention of the pulley

A.D.

105 invention of the papermaking process

400s discovery that the body can repair itself

400s discovery that diseases have natural causes

1100 invention of the magnetic compass

1100 invention of the rocket

1250 invention of the cannon

1440 invention of the printing press

1500s creation of first maps

1593 invention of the thermometer

1600s development of the microscope

1628 discovery that blood circulates

1687 formulation of the law of gravity

1687 invention of steam engine

1700 beginning of modern chemistry

1700s discovery of oxygen

1700s establishment of system for naming and classifying plants and animals

1752 discovery that lightning is electricity

1777 discovery that oxygen is needed to burn matter

1783 invention of the parachute

1802 discovery of atoms

1822 invention of the camera

1825 invention of the electromagnet

1830s proposal of the idea that all living things are made of cells

1831 discovery of electromagnetism

1837 invention of the telegraph

1846 invention of the sewing machine

1846 discovery of anesthetics

1849 invention of the safety pin

1852 invention of the passenger elevator

1858 invention of the Bunsen burner

1858 invention of the electric refrigerator

1866 invention of dynamite

1867 invention of the washing machine

1867 invention of the typewriter

1869 publication of the first periodic table of elements

1870 invention of margarine

1870 invention of electric light bulb

1876 invention of the telephone
1877 invention of the sound recording
1879 invention of the cash register
1800s discovery of the laws of heredity
1800s discovery that microorganisms cause disease
1885 invention of the motorcycle
1892 invention of the diesel engine
1892 invention of the zipper
1895 demonstration of the first radio
1898 invention of the submarine
1898 discovery of radium
1899 invention of the tape recorder
1901 discovery of X-rays
1902 invention of the air conditioner
1903 first flight of an airplane
1905 publishing of *Theory of Relativity*
1910 invention of neon lights
1917 invention of the helicopter
1925 invention of the television
1928 discovery of penicillin
1930s discovery of atomic energy
1930 invention of the jet engine
1930 discovery of Pluto, the 9th planet
1935 invention of nylon *(first synthetic fiber)*
1938 invention of the ballpoint pen
1940 invention of the photocopier
1947 first breaking of the sound barrier
1953 building of a model of DNA
1953 broadcast of first color TV program
1954 building of first nuclear submarine
1956 first videotape recording
1957 launching of *Sputnik*, the first human-made satellite to orbit Earth
1959 invention of the microchip
1960 invention of the laser
1965 invention of the holograph
1969 broadcast of first moonwalk
1971 invention of the microprocessor
1979 invention of the CD
1981 launch of the first space shuttle
1983 isolation of virus that causes AIDS
1990 launch of Hubble Space Telescope
1991 arrival of the World Wide Web; widespread Internet use begins
1995 first full operation of the GPS
2000 mapping of the human genome
2001 discovery of a large celestial body in the solar system beyond Pluto *(Quaoar)*
2002 creation of first synthetic virus
2003 first photo of a gamma-ray burst
2004 discovery of true age of the galaxy: 13.6 billion years

Some Memorable Scientists

These are just a few of the hard-working, creative people who have made some astounding discoveries and accomplished some important work in science. In most cases, these people worked their whole lifetimes searching for answers. You can explore their lives and work more completely. Find out more about their work! Find out what other contributions they made to science.

Aristotle (300s B.C.) – known for his works on logic, metaphysics, ethics, and politics

Sara Josephine Baker (early 1900s) – invented safe infant clothing; found a way to make silver nitrate drops safe for babies' eyes

Alexander Graham Bell (late 1800s-early 1900s) – invented the telephone

George Washington Carver (early 1900s) – discovered over 300 products that could be made from peanuts (ex: oil, cheese, soap, and coffee)

Nicolaus Copernicus (around 1500) – demonstrated that the Sun is the center of the solar system

Martha Coston (mid to late 1800s) – developed and manufactured maritime signal flares

Marie & Pierre Curie (late 1800s-early 1900s) – known for their work on radioactivity and discovery of radioactive elements

Charles Darwin (1800s) – traced the origin of man; explained theories of evolution in *The Origin of Species by Means of Natural Selection*

Thomas Alva Edison (late 1800s-early 1900s) – developed the electric light bulb, phonograph, mimeograph machine, and motion pictures

Albert Einstein (early 1900s) – discovered that mass can be changed into energy and that energy can be changed into matter; also known for his *Theory of Relativity*

Alexander Fleming (early 1900s) – developed penicillin

Henry Ford (early 1900s) – built the first gasoline engine and the first automobile

Benjamin Franklin (mid 1700s) – proved that lightning is electricity

Galileo Galilei (early 1600s) – formulated the *Law of Falling Bodies* and wrote about acceleration, motion, and gravity; developed the first astronomical telescope and made many discoveries in astronomy

William Harvey (1600s) – showed how blood circulates through the human body

Heinrich Hertz (late 1880s) – discovered electromagnetic radiation

Hippocrates (300-400 B.C.) – founded the first school of medicine; known as the *Father of Medicine*

Edward Jenner (late 1900s) – founded the science of immunology by developing a vaccine to protect the body against smallpox

James Joule (1840s) – showed that heat is a form of energy

Stephanie Kwolek (early to mid 1900s) – developed Kevlar, (a fiber stronger than steel) used in bullet resistant clothing

Carolus Linnaeus (mid 1700s) – a Swedish naturalist and physicist who devised a systematic method for classifying plants and animals

Barbara McClintock (mid to late 1900s) – Nobel Prize winner (1983) who did pioneer work with genes, studying cell structure in corn kernels

Gregor Johann Mendel (mid 1800s) – founded genetics through his work with recessive and dominant characteristics of plants

Dmitry Ivanovich Mendeleyev (late 1800s) – developed the periodic table for classification of the elements in 1869

Sir Isaac Newton (1600s) – discovered that the force of gravity depends upon the amount of matter in bodies and the distances between the bodies; formulated the laws of gravity and motion

Louis Pasteur (mid 1800s) – developed a method for destroying disease-producing bacteria and for checking the activity of fermentative bacteria (pasteurization)

Jonas Salk (mid 1900s) – an American physician and bacteriologist who developed a vaccine to prevent polio

James D. Watson and Francis H. C. Crick (mid 1900s) – created a model of the molecular structure of DNA; won the Nobel Prize in 1962

James Watt (late 1700s) – invented the modern steam engine. The *watt*, a measure of electrical power, was named after him.

Eli Whitney (late 1700s-early 1800s) – invented the cotton gin and developed a faster way to make manufactured goods

Orville and Wilbur Wright (early 1900s) – succeeded at the first controlled and sustained airplane flight at Kitty Hawk, North Carolina on December 17, 1903

A Remarkable Woman

Navy Rear Admiral Grace Hopper, known for her advancements in computer technology, is often called the *Mother of the Computer.* She is probably the only woman scientist to have a ship named after her, the ***USS Grace Hopper DDG 70.*** She is also famous for saying, *"It is easier to ask forgiveness than it is to get permission."*

I'm going to be a scientist myself, someday.

Looking for Answers

Scientists have a serious, consistent way of looking for answers about the world. They follow some steps that other scientists can repeat.

A good scientist follows the steps but tries to think outside the box.

Scientific Inquiry Steps

1. **Observe**
2. **Form a question**
3. **Hypothesize**
4. **Plan an investigation**
5. **Gather data**
6. **Analyze & summarize data**
7. **Give explanations**
8. **Communicate results**

How Do Scientists Work?

They . . .

. . . observe carefully

. . . take notes

. . . ask questions

. . . look at (and read) what already is known

. . . examine and use knowledge from many branches of science

. . . compare their questions and answers to knowledge already available

. . . form hypotheses (careful, informed guesses) about what could be true

. . . plan investigations that can test a hypothesis

. . . follow clear steps and processes of scientific inquiry

. . . control the variables that are not being examined in their investigation

. . . use tools and instruments to collect and classify data

. . . use tools and instruments to organize and summarize data

. . . use math to measure, calculate, and communicate findings

. . . keep careful records of their results

. . . analyze and describe the results completely

. . . use logic to understand relationships and draw conclusions

. . . draw conclusions and develop explanations for their findings

. . . provide evidence to support their conclusions

. . . review and repeat their investigations

. . . make their methods of study and results public

. . . review and repeat studies done by other scientists

. . . keep accurate, understandable records

. . . pay attention to new questions raised by their investigations

Kate's Inquiry

1. Observe

Kate picked up a rubber comb, used her sweater to brush away the dust and lint on it, and started to comb her hair. She noticed that the comb crackled like electricity and stood on end to meet the comb. It was almost as if the comb was a magnet!

2. Question

Will my comb attract other substances besides hair?

3. Hypothesize

Kate guessed that her comb would act as a magnet to attract other lightweight substances.

4. Investigate

Kate found several lightweight substances: pepper, puffed cereal, and uncooked minute rice. She sprinkled these on plates. She rubbed the comb on her sweater and held it close to each substance.

5. Gather Data

Look! The pepper is jumping toward the comb.

The rice jumps, too.

So does the cereal.

6. Analyze and Summarize

All three substances jumped up to meet her comb. This happened only after she "charged" the comb by rubbing it against the wool.

7. Explain Results - 8. Communicate Results

Rubbing the comb on the wool created static electricity. Static electricity is caused when tiny electrons jump around. Rubbing the comb on the wool caused the electrons to get in motion. Once the comb was "charged", it attracted other objects that have an opposite charge, or **no** charge at all.

I made a drawing to show my results.

Science Processes

As scientists work and wonder, they use certain processes (or skills of *doing*). You may not be a professional scientist, but if you are going to try to uncover facts, answer questions, or solve mysteries of the natural world, you will need to be good at using these same processes.

Observing – using the senses to obtain information

If you pay attention to what your eyes, ears, nose, tongue, and skin are telling you, you might notice such things as . . .

. . . *the heat in your palms when you rub your hands together quickly*

. . . *the strange smell that rises when you smash a cooked egg yolk*

. . . *the change in pitch of a siren sound as the speeding ambulance passes you*

. . . *the sediment that collects in the bottom of your glass of iced tea*

. . . *the difference between the taste (and feel) of a green banana and a very ripe banana*

Comparing – observing how things are alike or different

As you observe, keep a ***comparing*** attitude. Constantly match up one event against another. For instance, you might notice . . .

. . . *that all the body joints are located at places where bones come together*

. . . *that all body joints allow movement of some kind*

. . . *that some joints rotate, while others bend like a hinge, swivel, or twist*

Questioning – raising uncertainty or making a statement that seeks an answer

Observations often raise questions. For instance . . .

. . . *Why does a cooked egg spin smoothly while a raw egg wobbles all over?*

. . . *Will a magnet attract all objects made of metal?*

. . . *Does the color of a glass affect the rate at which the ice melts?*

Hypothesizing – tentatively accepting an explanation as the basis for further investigation

Observations can lead to a smart guess about what might happen in an investigation. A hypothesis should be made carefully. It must be a statement that can be tested in an investigation or experiment.

Can be tested:

. . . *The speed of a ball rolling down an incline will increase as the incline angle increases.*

. . . *Hot water freezes faster than cold water.*

Can't be tested:

. . . *Salt and water form a solution faster than any other two substances.*

Experimenting – testing under controlled conditions

Controlling Variables – managing factors that may influence an experiment

An experiment should be carefully planned and controlled. It is important to manage any factors which might influence the outcome. Only the factor being tested should be allowed to vary. The other factors (variables) should be controlled to remain the same. For instance . . .

Max does an experiment to find out which of several substances will melt the fastest. The ***variables*** are the five different frozen substances (water, butter, ice cream, raw hamburger meat, and mashed potatoes). Here's what does not vary. (These things are controlled.) He has used a cup to measure exactly the same amount (and same thickness) of each substance. They are all placed in identical containers and frozen solid for 24 hours. They are all removed from the freezer at the same time and set in the same location in a place of the same temperature.

I will check my frozen substances after 3 hours, 6 hours, and 12 hours to see which melts the fastest.
water
butter
ice cream
raw hamburger
mashed potatoes

Classifying – assigning objects or processes to a category based on a common characteristic

Events, objects, or processes belong to one or more groups or classes. When you group something, make sure it shares the characteristics of the other items in the group. Here are some examples . . .

. . . *Vertebrates can be divided into classes: fish, reptiles, mammals, amphibians, and birds.*

. . . *Copper, neon, wood, water, nitrogen, iron, and sand are all substances. But copper, neon, nitrogen and iron can be classified as elements. The others are not elements. Wood, water, and sand are compounds.*

Defining Operationally – listing the characteristics or behaviors by which something is defined

Here is an operational definition of a liquid . . .

A liquid is a substance that can be poured. It has no shape of its own, but takes on the shape of whatever container into which it is placed.

Formulating Models – **devising a concrete representation to illustrate an abstract idea or relationship, or to show something that cannot be seen easily**

A model can be a physical (3-dimensional) creation, a picture or diagram, or a mathematical formula. Models can be built, drawn, designed on a computer, or written.

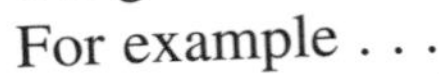

For example . . .

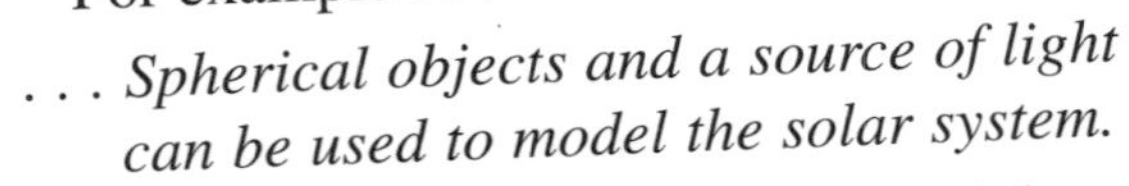

. . . *Spherical objects and a source of light can be used to model the solar system.*

. . . **H_2O** *is a mathematical model of the compound water.*

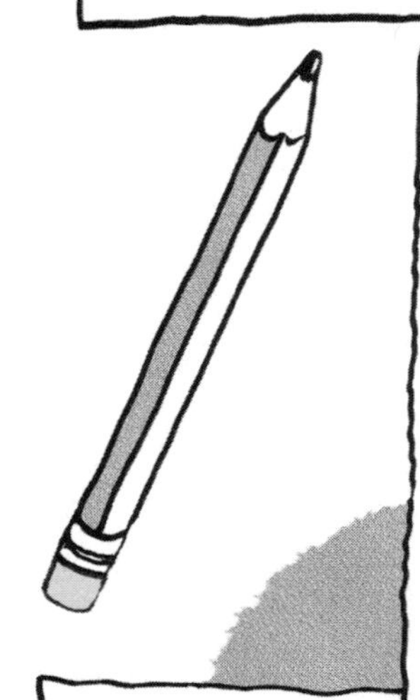

Using Math – **using numbers to count, measure, or otherwise give quantities of data**

Science and math are close friends. Math is needed for just about every science question or investigation. For instance, you'll use math to . . .

. . . *count the number of fruit flies hatched since yesterday.*

. . . *measure the distance you tossed each of those eggs.*

. . . *find the difference in the height of two plants you're growing.*

. . . *calculate the speed of a pendulum swing.*

Summarizing & Recording Data – **systematically expressing the results of collected data and storing it in an organized fashion**

The data found in an investigation must be recorded if it is to be used. The results should be organized in such a way that they make sense as a part of the record. The record might be written, typed into a computer, recorded on an audio or videotape, or represented some other way.

Why do I have to keep good records? Isn't **doing** it enough?

BECAUSE . . .

Others can learn from your record.

Recording shows what you found out.

A good record makes it possible for others to follow your steps and repeat your investigation.

Recording shows others what you did.

Records help you remember and explain what you did.

Interpreting Data – finding patterns or relationships in a set of data

When you look at information gathered from an investigation, search for patterns in the data and look for relationships among the events.

For example . . .

I notice that every one of the mixtures that bubbled had two ingredients in common: carbon dioxide and baking soda.

The only pieces of fruit that did not turn brown were the ones dipped in orange juice.

Inferring – deriving a conclusion from available evidence

So what does the evidence (data) tell you? When your investigation is finished, you will need to come to some conclusion about what you have found.

I poured some boiling water into this can. As the can cooled, it collapsed. I infer that the air pressure outside the can was greater than the pressure inside the can.

Predicting – foretelling from previous information

Sometimes the results of an investigation give you data that make it possible for you to predict events or results you haven't seen.

The invisible ink shows up when I hold the paper near a light bulb. I predict that this ink will turn brown if I put it near any heat source.

Communicating – exchanging information

New information or results from an investigation don't do much good if they are not shared. Practice ways of communicating what you've learned. These are just a few ways . . .

drawing or diagram	model	graph	written log
videotape	article	poster	book
typed record	tape recording	e-mail	table or chart

Safe Practices in the Science Lab

Investigating science questions can be fun, amazing, or fascinating. It also can be hazardous. Safe science is a serious matter. The equipment, substances, and processes all need to be handled with great care. If you follow good practices for science safety, you'll discover a lot more, and have plenty of excitement without danger.

1. Always get permission from a teacher or adult before starting an investigation on your own.
2. Review the lab procedure before you start, so you know what you'll be doing.
3. Keep a fire blanket, fire extinguisher, first aid kit, and eyewash kit nearby.
4. Use the safety equipment. Wear goggles and an apron during experiments.
5. Use only lab equipment. Don't do experiments with utensils you are using for something else.
6. Tie back loose hair and clothing when working in the lab.
7. Be careful to keep excess material away from flames.
8. Always slant a test tube away from yourself when you are heating a substance.
9. Never inhale or taste any substances or chemicals.
10. Never eat out of containers in the lab. (In fact, don't eat in the lab at all!)
11. Handle glass and all equipment as if it were hot or dangerous.
12. Be especially cautious around chemicals, heat sources, fumes, or electrical equipment.
13. If you spill anything on your clothes or skin, wash it off with lots of water immediately.
14. If anything gets in your eye, wash it with lots of water—immediately.
15. If your clothes catch on fire, DO NOT RUN. Smother a fire under a coat or fire blanket.
16. Flush any burns with cold water immediately.
17. Tell the teacher about all spills, splashes, accidents or injuries—right away.
18. Return all substances to their original containers. Close containers tightly.
19. Do not wash any substances down the sink. Dispose of them as you are directed.
20. Before you leave the lab, make sure all sources of heat are turned off and all electrical devices are disconnected.

TOP 10 BAD IDEAS
FOR SCIENCE LAB BEHAVIOR
10.
yuck
Number 10: Taste the chemicals to see if you can identify them.
9.
Number 9: Insist to yourself (and everyone else) that you can see better without your safety goggles.
7.
rub
oops!
Number 8: Don't worry if your charming new hairdo gets in the way of your experiment.
8.
Number 7: Wipe up all chemical spills with your shirt sleeve.
Number 6: Take good strong whiffs of the chemicals you are heating or mixing.
6.
sniff
Number 5: Always slant the test tube toward you when you are heating or mixing substances.
5.
Number 4: Use an empty test tube as a goblet for drinking lemonade.
4.
3.
pop
pop
pop
Number 3: Get a brilliant notion to use your Bunsen burner for popping corn.
2.
Number 2: Pour leftover chemicals down the sink.
1.
The Number 1 Bad Idea: If your clothes catch on fire - RUN!

The Big Ideas of Science

If you study science, you will keep running into these big ideas. No matter what specific topic, idea, or field you dabble in, these will pop up. That's because there are some basic underlying concepts (or big, broad ideas) that relate to or explain the natural world.

I'd like to learn more about the order of the moon phases.
I'd like to learn more about how to order some French fries.

Systems

A system is an organized group of related components or objects that form a whole and work together as a whole to perform one or more functions. Each system has components and boundaries. A change in one component of a system affects the whole system. Most systems function with an input and an output of materials and energy. Most systems are related or connected to other systems.

Examples:

- an ecosystem
- the solar system
- a digestive system
- any organism
- a river system
- an electrical circuit
- a cyclone
- a pulley

Organization

Organization is an arrangement of independent items, objects, organisms, units of matter, or systems joined into one whole system or structure.

Examples:

- Living things are organized into several kingdoms.
- The planets are organized in a particular order from the Sun.
- Elements are organized by the structure of their atoms.

Order

Order is the predictable behavior of objects, units of matter, events, organisms, or systems.

Examples:

- When an animal cell is fertilized, it can be expected to start dividing.
- Year after year on Earth, fall follows summer and spring follows winter.
- As a substance gets cooler, the molecules will move more slowly and stay closer together.

Cycle

A cycle is a series of events or operations that occur regularly and usually lead back to the starting point.

Examples:

- rock cycle
- life cycle of stars
- water cycle
- nitrogen cycle
- oxygen-carbon dioxide cycle
- life cycle of every living organism
- movement of planets
- rising and falling and changing of tides
- phases of the Moon

Change

Change is the process of becoming different. Change is a key part of the processes in the natural world. Changes occur in properties, motion, positions, form, and function of objects and materials. Changes occur when materials, objects, organisms, and systems interact with each other. Changes can be described, measured, and quantified.

Examples:
- weathering of rocks by wind
- growth of an infant into an adult
- lack of food causing an animal population to shrink
- rivers moving sediment downstream

The size of my nose is changing.
My freckles are still here every morning. They are a constant in my life.

Constancy

Constancy is the opposite of change—a state characterized by the lack of change. While most things are changing, there are some properties and processes that remain the same.

Examples:
- The total mass plus energy in the universe is always the same.
- The temperature of boiling water stays the same, even if more heat is added to the water.

Evolution

Evolution is a series of changes that causes the form or function of an object, organism, or system to be what it currently is. Things in the natural and manmade world change over time. The present forms and functions of systems, organisms, objects, and materials have arisen from the past.

Examples:
- The universe as a system is constantly changing with the birth and death of stars.
- After repeated exposure to certain antibiotics, some bacteria evolve to become resistant to the antibiotic so they are no longer affected by it.

Equilibrium

Equilibrium is the state in which equal forces occur in opposite directions and offset or balance each other. Most systems and interacting units of matter tend toward a state of balance—a steady state in which energy is uniformly distributed.

Examples:
- When the body temperature gets too high, the body sweats to help regulate the temperature back to normal so the body doesn't overheat.
- When there is too much pressure in an area beneath Earth's surface, a volcano may erupt to release and equalize the pressure.

Form & Function

The shape or structure (form) of an organism, object, or system is often related to its operation (function). Frequently, the function of something is very dependent upon its form.

Examples:

- The wall of a plant cell is very sturdy. *(form)*
  This allows the plant cells to form sturdy stems or trunks. *(function)*
- The snail has a sticky, muscular foot. *(form)*
  This foot helps the snail to hold on to surfaces and pull its way along. *(function)*

I have webbed feet. The structure (form) of my feet helps me to push against water (function). This makes me a good swimmer.

Cause & Effect

A cause is anything that brings about a result. An effect is the result (the event or situation or behavior) that follows from the cause.

Examples:

- Cause: An infection breaks out in the body.
  Effect: The body's defenses start to fight the infection.
- Cause: A predator comes close to a rattlesnake.
  Effect: The snake signals a warning by shaking its tail.

Energy & Matter

Energy and matter are closely related. They are constantly interacting with each other. This means that they affect one another or work together. Energy can move or change matter. Matter can be changed into energy. Energy can be transferred to matter.

Examples:

- The heat of sunlight causes ice cubes to melt.
- When wood burns, it is changed to a different substance. Heat is released.
- The energy of moving water moves the sand along the shore.

Force & Motion

Force and motion have a close relationship. An object changes position, direction, or speed when a force acts on it.

Examples:

- Muscle power pushing a swing moves the swing up in the air.
- A girl in a rowboat pulls her oars backwards in the water. The boat moves forward.
- A pitcher pitches a baseball to the batter. The batter swings. The ball collides with the bat, and changes direction, flying over the infield.

GET SHARP

in

SPACE SCIENCE

There have been more rodents in space than frogs, you know.
SPACE Science

The Solar System

What a strange and wonderful neighborhood: one star, nine planets, many moons, thousands of asteroids and comets, and billions of other bits of rock all moving around! This is the place we call home—our solar system. Most of the solar system is empty space, even with all these billions of things moving within it.

The centerpiece of the solar system is our Sun, a brilliant star. Planets orbit around the Sun. Many of these planets have moons that orbit around them. Between the orbits of Mars and Jupiter, thousands of asteroids orbit the Sun in an asteroid belt.

Comparative Sizes of the Planets (Distances shown are not proportional)

Distance from the Sun in Astronomical Units

The Sun's great size gives it strong **gravitational pull**. This is the force that keeps the solar system together and keeps the planets revolving around the Sun.

Each planet revolves around the Sun in a path called an **orbit**. These orbits are elliptical in shape. The orbits are just about in the same plane, except for Pluto's orbit, which varies by about 17 degrees.

Since the orbits are elliptical, each planet is not always the same distance from the Sun as it travels. The **perihelion** is the point in the orbit where the body is closest to the Sun. The **aphelion** is the point where the object is farthest from the Sun.

In 2001, astronomers discovered a large celestial body in the solar system beyond Pluto. It is named Quaoar. This was the largest object discovered since the 1930 discovery of Pluto.

In 2003, another body was discovered. Named Sedna, it is the farthest natural object ever observed in the solar system. Neither body has yet been called a planet.

An astronomical unit is the distance from Earth to the Sun (92.9 million miles).

The Sun

The hot ball of gas that lies at the center of our solar system is called the Sun. It's just one of millions of stars in the Milky Way galaxy. Although it is a huge, brilliant star, it is not unusual. It is, however, the most important star in our lives.

The Sun is a restless flame that is always changing. It bubbles and spurts and hurls hot gases out millions of miles beyond its surface. The Sun is a nuclear furnace. In its core, nuclear fusion changes hydrogen to helium. These reactions produce energy, which radiates outward from the core. As the energy travels outward from the center, it is absorbed in the layers. When it is finally transmitted at the surface, the temperatures are considerably lower.

Sun's Vital Statistics

Temperature at the center:
about 30 million° F

Temperature at the surface:
about 11,000° F

Diameter:
almost 900,000 miles

Age:
thought to be about 4.5 billion years old

Density at center:
over 150 times the density of water

Distance from Earth:
93 million miles

Present composition:
75% hydrogen & 25% helium

Also of interest:

- The Sun's light takes 8 minutes to reach Earth.
- The composition of the Sun changes over time as hydrogen is converted to helium in the core of the Sun.
- The Sun converts four million tons of gases into energy every second.

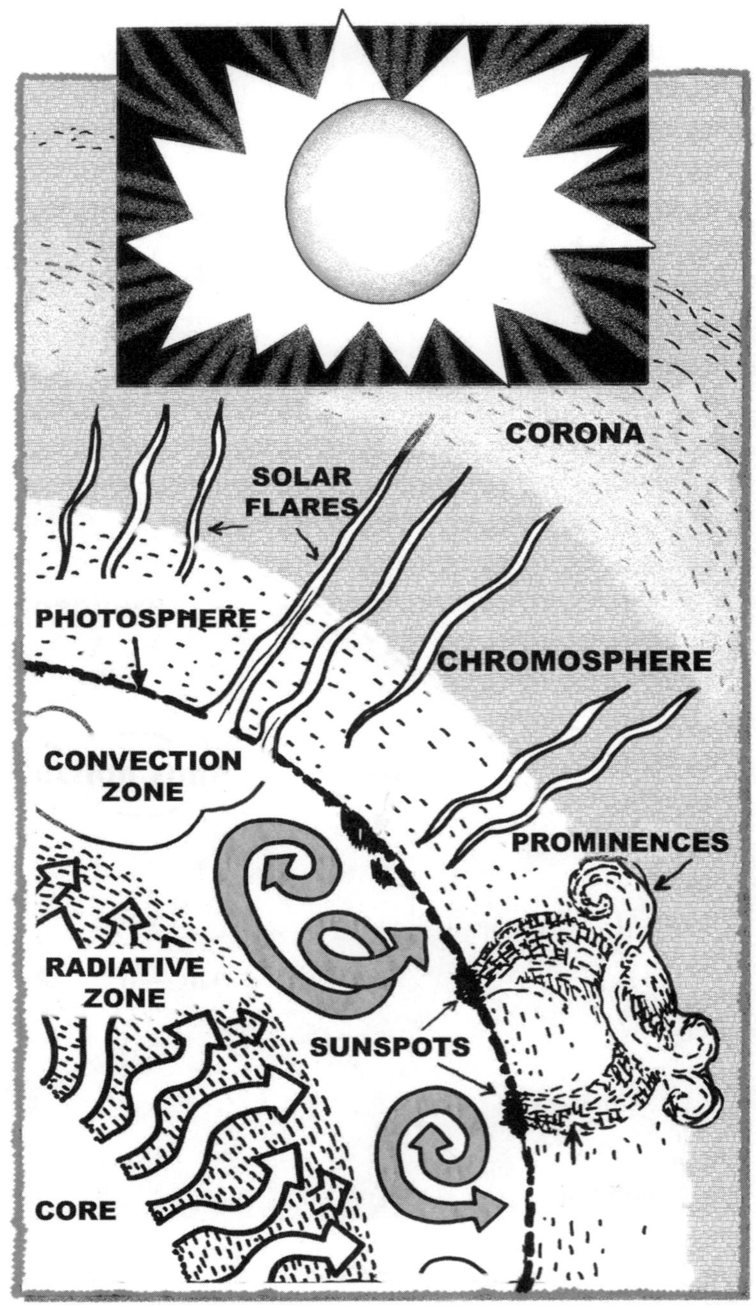

The Sun's Composition & Structure

Core – Heat and light are produced in this center of the Sun by nuclear fusion reactions.

Radioactive Zone – Energy from the reactions push out through this area, flowing in waves.

Convection Zone – The waves reach a place where they no longer have the strength to keep pushing forward. In this region, the waves churn around, and the heat travels on by convection.

Photosphere – This is the name given to the visible surface of the Sun. It emits the visible radiation that we can see.

Chromosphere – This bright red layer of gas extends about 6,000 kilometers above the photosphere. This layer can be seen only during an eclipse.

Prominences – These "fingers" of flame are glowing gas that rises from the Sun's surface in surges and shoots outward from the chromosphere.

Corona – This is the outer "atmosphere" of the Sun—a transparent region beyond the chromosphere that extends millions of miles into space. This is visible only during eclipses.

Solar Flares – These are sudden increases in brightness in the chromosphere. These occur near sunspots. Electrons and protons leap out from these areas at high speeds, sometimes reaching and affecting Earth's atmosphere.

Sunspots – These are relatively cooler areas of the Sun that look dark by comparison. Sunspots develop and may last for a day or as long as several months. They can be thousands of miles in diameter.

Solar Wind – The Sun gives off a low-density stream of charged particles known as the solar wind. The wind can affect the aurora borealis, the tails of comets, and the travels of spacecraft.

NEVER!

Never look directly at the Sun, even through a telescope or binoculars, even on a cloudy day. Radiation from the Sun is very dangerous and can cause blindness.

See page 85 for a suggested way to look at the Sun.

Sometimes, solar flares disturb gases in the upper atmosphere of Earth, causing them to radiate some spectacular displays of colored lights in the sky.

These lights are called the **aurora borealis** in the Northern Hemisphere and the **aurora australias** in the Southern Hemisphere.

Planetary Particulars

A ***planet*** is a large body that orbits a star. In our solar system, we know of nine planets that orbit our sun.

The four ***inner planets*** (Mercury, Venus, Earth, Mars) are referred to as the ***terrestrial*** (or rocky) ***planets***. They are composed mostly of rock and metal.

In general, these planets have slow rotation, solid surfaces, high densities, few satellites (or moons), and no rings.

Get Sharp #1
With the exception of Earth, the planets are named after Greek and Roman gods.

The Inner Planets

	Mercury	Venus	Earth	Mars
Average distance from the Sun (in miles)	36 million	67 million	93 million	142 million
Diameter at the equator (in miles)	3,049	7,565	7,926	4,220
Mass (times Earth)	.06	.82	1 (6.6 sextillion tons)	.11
Length of Day/Night (Rotation time in Earth time)	176 days	117 days	23 hours, 56 minutes	24 hours, 37 minutes
Revolution time (In Earth time)	88 days	225 days	365 days, 5 hours	687 days
Surface Gravity (Earth = 1.0)	0.38	0.9	1.0	0.38
Average Temperature or Temperature Range	–300° to 800°F	900°F	–136° to 128°F	–116° to 32°F
Number of Satellites or Moons	0	0	1	2

Four of the five ***outer planets*** (Jupiter, Saturn, Uranus, Neptune, and Pluto) are referred to as the ***gas planets***. They are composed mostly of gases and some liquid. In general these planets have rapid rotation, low densities, rings, and many satellites (or moons). Pluto's composition, which is mostly dust and ice, is unlike any of the other planets.

Do you know what keeps the Sun's gravity from pulling all the planets right up next to it?
Here is the answer: Each planet has its own gravity. The two forces balance just right to keep each planet in its orbit.
Two other bodies have been discovered in the solar system beyond Pluto: Quaoar and Sedna. To this date, neither has been called a planet.

Did You Know?

An astronaut who weighs 200 pounds on Earth would have these weights on other planets:

Mercury	86 lb.
Venus	180 lb.
Mars	86 lb.
Jupiter	540 lb.
Saturn	240 lb.
Uranus	186 lb.
Neptune	200 lb.
Pluto	6 lb.

The Outer Planets

	Jupiter	Saturn	Uranus	Neptune	Pluto
Average distance from the Sun (in miles)	484 million	887 million	1.8 billion	2.8 billion	3.7 billion
Diameter at the equator (in miles)	89,500	75,000	31,135	30,938	1,875
Mass (times Earth)	317.9	95.2	14.54	17.2	.002
Length of Day/Night (Rotation time in Earth time)	9 hours, 50 minutes	10 hours, 14 minutes	17 hours, 14 minutes	17 hours, 6 minutes	6 days, 9 hours
Revolution time (In Earth time)	11.9 years	29.5 years	84 years	165 years	248 years
Surface Gravity (Earth = 1.0)	2.7	1.2	0.93	1.5	0.03
Average Temperature or Temperature Range	–238° F	–274°F	–328°F	–346°F	–380°F
Number of Satellites or Moons	16	18	15	8	1

The Inner Planets—Vital Statistics

Mercury – The Speediest Planet

Mercury travels at 29.7 miles per second in its orbit.

- closest to the Sun; the first planet; second smallest planet
- dark in color; rocky planet
- orbit is one of the most elliptical of all planets
- fastest planet in the solar system
- shortest year (revolution time)
- longest day (rotation time)
- no moons or other satellites
- extremely thin atmosphere
- strong magnetic field
- heavily cratered surface and wrinkled crust
- extreme hot and cold temperatures
- second densest planet

Also of interest:

- A large crater called the *Caloris Basin* is one of the largest surface features on Mercury. It was probably created by an impact with some other body in the solar system.

Mercury is the god of commerce, travel, and thieves. Mercury is also the messenger of the Roman gods.
Venus is the Roman god of beauty.

Venus – The Visible Planet

Venus is usually visible from Earth without a telescope.

- second planet from the Sun; sixth largest planet
- nearly circular orbit
- dense carbon dioxide atmosphere
- hotter than Mercury
- slow, retrograde rotation
- mostly rock, with surface of mountains, volcanoes, large craters, and lava plains
- extremely dense atmosphere
- no moons or other satellites
- atmosphere traps sun's radiant heat

Also of interest:

- Sulfuric acid drips from the clouds, giving the planet an orange color.
- Venus may have had water, but it all boiled away from the heat.

Earth is the only planet whose name does not come from Greek or Roman mythology. It is of Germanic and Old English origin.
Mars is the Roman god of war.

Earth – The Home Planet

Earth is the only planet known to have an atmosphere that can sustain life.

- third planet from the Sun; fifth largest planet
- densest major body in the solar system
- solid inner iron core and rocky crust; crust varies in thickness
- semi-fluid outer core and mantle layers
- plants release oxygen into the atmosphere
- atmosphere is 78% nitrogen, 21% oxygen, 1% other gases
- ozone layer absorbs dangerous ultraviolet rays from the sun
- only planet with liquid water on surface
- one natural satellite

Also of interest:

- Humans have placed thousands of artificial satellites into orbit around Earth.
- Earth could not be fully explored until spacecraft could be launched to see it from space.

Mars – The Red Planet

A rusty surface coating is blown around by great storms, making the planet look red.

- fourth planet from the Sun; seventh largest planet
- highest mountains and deepest valleys in solar system
- permanent ice caps at the poles of frozen carbon dioxide
- winds raise huge dust storms at the surface
- thin atmosphere, mostly of carbon dioxide
- covered with a thick cloud layer
- surface of craters, mountains, valleys, volcanoes
- two natural satellites
- highly interesting and varied terrain

Also of interest:

- Many scientists believe there may be traces of liquid water on Mars.
- Mars is a bright planet, visible on Earth when it is in the nighttime sky.

The Outer Planets - Vital Statistics

Jupiter – The Giant Planet

All the other planets could be squeezed into Jupiter.

- largest planet; fifth planet from the Sun
- fastest-spinning planet (day lasts less than 10 hours)
- fourth brightest object in the sky
- a gas planet—90% hydrogen, 10% helium
- three layers of icy clouds
- radiates heat from an internal heat source
- strong magnetic field
- has a great red spot *(probably caused by a violent storm)*
- dark rings, probably of rocky particles
- 16 satellites that have been named
- largest moon in the solar system *(Ganymede)*

Also of interest:

- The fast rotation pulls the planet slightly out of shape with a bulge at the equator.
- High winds blowing in opposite directions cause light and dark colored bands that are features of Jupiter's vividly colored appearance.

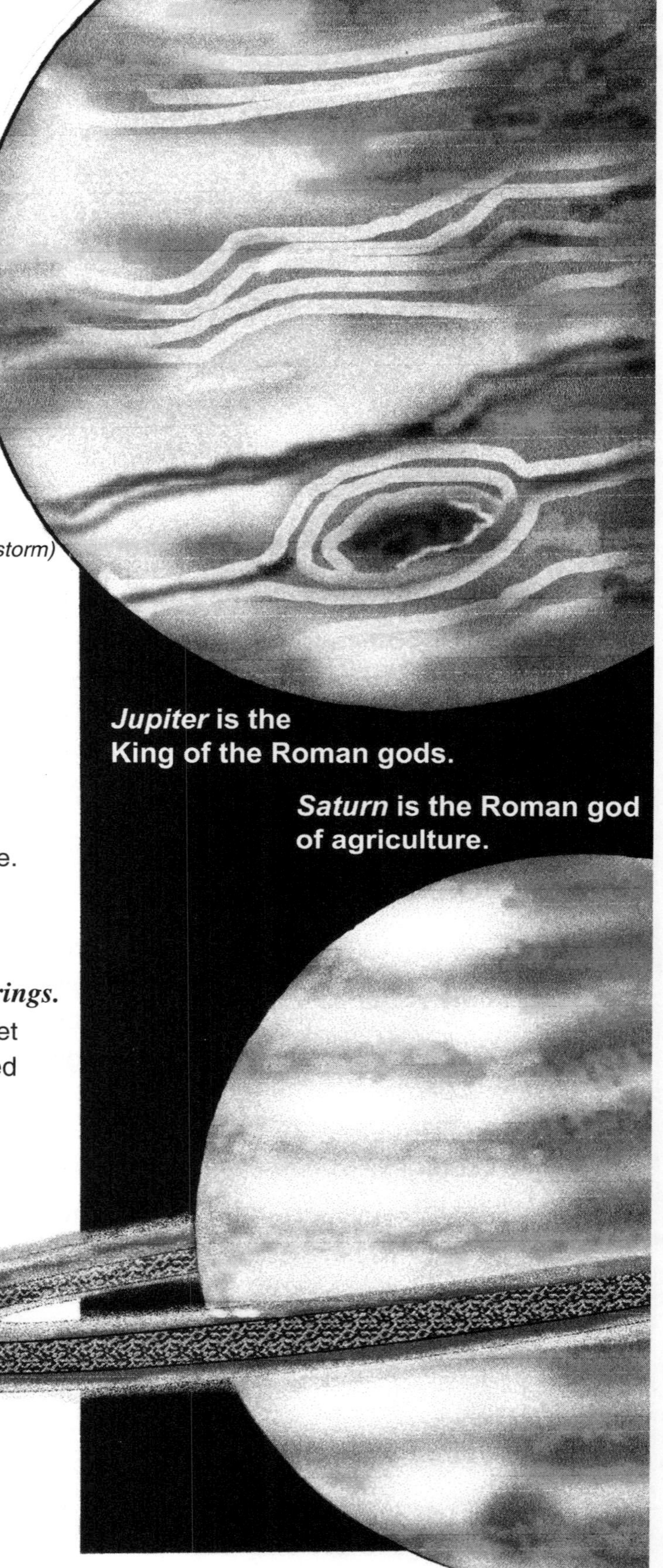

Jupiter is the King of the Roman gods.
Saturn is the Roman god of agriculture.

Saturn – The Ringed Planet

It has the largest and most complicated system of rings.

- sixth planet from the Sun; second largest planet
- over 1000 rings extending 46,000 miles; formed from icy particles
- radiates heat from an internal heat source
- composed of 75% hydrogen, 25% helium
- a great brown spot near the north pole
- flattened from fast rotation
- least dense planet
- strong magnetic field
- 18 named satellites

Also of interest:

- Saturn is less dense than water. It could float.
- Every 30 years or so, violent storms cause bright white spots on the surface.
- The origin of Saturn's rings is unknown.

Uranus – The Tilted Planet

The axis of Uranus is tilted so far that the planet virtually orbits on its side.

- seventh planet from the Sun; third largest planet
- discovered by accident in 1871
- retrograde rotation
- 11 known dark rings of rock, dust, and ice
- 15 known satellites, named for Shakespearean characters

Also of interest:

- Each pole stays in darkness for 40 years *(because of the planet's tilt).*
- Methane in the atmosphere absorbs red light, causing the planet to appear blue.

Neptune — The Windy Planet

Neptune has the fiercest winds in the solar system—blowing up to 1,400 miles per hour.

- eighth planet from the Sun; fourth largest planet
- gaseous planet; radiates heat from an internal heat source
- brilliant blue methane atmosphere
- three thin, dark rings around the planet
- 8 known moons

Also of interest:

- Great Dark Spot changes size and shape every 16 days.
- Every 250 years, Neptune is the outermost planet when Pluto comes closer to the Sun for 20 years at a time.

Pluto — The Tiny Planet

Pluto is smaller than Earth's moon, and by far the tiniest planet in the solar system.

- farthest planet from the Sun
- most recently-discovered planet
- the most elliptical orbit
- seemingly a snowball of frozen gases
- thin atmosphere
- severely tilted axis (over 62°)

Also of interest:

- Pluto is the only planet never visited by a spacecraft. It is too far away even to be seen clearly by telescopes.
- Pluto is so similar in size to its moon, Charon, that many astronomers consider them a double planet.

Uranus is a Greek deity of Heaven.

Neptune is the god of the sea in Roman mythology.

Pluto is the Greek god of the underworld.

Other Solar System Objects

The solar system contains thousands of other interesting objects in addition to the sun and the nine planets. Most of them smaller than the planets. In general, the objects are classified as ***asteroids*** or ***comets***. These bodies are sometimes called ***minor planets*** or ***planetoids***. And there's even more stuff out there in that space that seems empty. There are billions of particles of dust (much of it microscopic) floating around. This is often referred to as ***space dust***.

Asteroids

Asteroids are small, dense, rocky objects that orbit the Sun. They are fragments of material similar to the inner planets, ranging in size from very tiny to miles across. Most asteroids orbit in a belt of 10,000 or more between Mars and Jupiter. The largest asteroids are named *(for example: Ceres, Psyche, Juno, Davida)*.

The largest asteroid, Ceres, has a diameter of 620 miles.

Meteoroids

Very small asteroids orbiting the sun are called ***meteoroids*** to distinguish them from larger asteroids. Some meteoroids are particles thrown off from the nucleus of comets.

Meteors, Shooting Stars, & Fireballs

Millions of meteors enter Earth's atmosphere every day. When this happens, the air friction heats the meteoroid, creating a glowing trail of gases. When a streak of a meteoroid is visible in the sky, it is called a ***meteor***. The streak of light flashing through the sky is quite spectacular, and meteors are often called ***shooting stars*** or ***falling stars***. An unusually bright meteor is called a ***fireball***.

Meteor Showers

When Earth passes through a large group of meteoroids, many may hit Earth's atmosphere. This can cause many meteors to be seen at once, resulting in a ***meteor shower***.

Meteorites

Generally, meteors burn up in Earth's atmosphere. When a piece of a meteor survives the trip through the atmosphere and collides with Earth's surface, it is called a ***meteorite***.

The asteroid with the orbit closest to Earth is Hermes. It comes within about 500,000 miles of Earth.

About 44,090 tons of cosmic dust falls on Earth each year.

The largest meteorite ever found fell in Namibia. It weighed about 60 metric tons.

By the way, I wouldn't try to bathe in a meteorite shower, if I were you!

Comets

A *comet* is a large clump of ice, dust, and frozen gases that travels around the Sun in a long orbit. As the body nears the Sun, some of the ice melts or vaporizes. Streams of gases and particles fly away from the main body of the comet and create a spectacular tail that shines in the sunlight and stretches for miles.

A comet is composed of different parts:

The **nucleus** is the mostly solid part of the comet. This consists of ice and gas with solid particles and dust.

The **coma** is a thick cloud of water, carbon dioxide, and other gases surrounding the nucleus. Though a nucleus may be only a few miles across, a coma can be wider than a planet.

The **tail**, made of dust and gas, is the most spectacular part of a comet. It can trail along behind the comet for millions of miles.

The **orbit** of a comet is an eccentric path that goes far beyond the farthest planet and back around the Sun. The orbit of a comet usually is not in the same plane as the planets and Sun, but is at an angle to the orbits of the planets. Orbits vary greatly in length. Some comets take thousands of years to complete an orbit. Comet Encke completes its orbit in about three years.

Comets are often called dirty snowballs or icy mudballs.

The most famous comet is Comet Halley. It reappears every 76 years. Its last appearance was in 1985 and 1986.

A comet can age. Each time a comet orbits the Sun, it loses some of its ice and dust. Eventually, a comet may lose its entire nucleus. It may break up into a bunch of asteroids, or turn into a group of meteors.

In 1997, Comet Hale-Bopp came within 122 million miles of Earth. The bright comet, with its large amounts of dust and gas, was easily visible from Earth.

Get Sharp Tip #2

Comets orbit the Sun. They are only visible when they get near the Sun.

Earth's Moon

Sometimes it is a huge, glowing ball. Sometimes it seems to disappear. It is beautiful, romantic, shadowy, eerie, and mysterious. The marks on its surface, the way it changes, and its shiny beauty have fascinated people for centuries. What is this Moon, our closest neighbor?

The Moon is the one ***natural satellite*** of our planet. The Moon is the brightest and biggest thing in the night sky. It is a rocky, cratered body orbiting Earth and shining from the reflected light of the Sun. Its topography is varied—with hills, mountain ranges, valleys, chains of craters, ridges, faults, and waterless "seas."

The Moon's Vital Statistics

Diameter: 2,160 miles
Distance from Earth: 221,000–252,000 miles
Orbit time: 27.3 days
Temperature: –274° F to 230° F
Day/Night: 29.5 days

Also of interest:

- surface covered with fine dust and rocky bits
- no wind; no weather; no atmosphere
- extreme variation in temperatures
- craters formed by collisions with other bodies

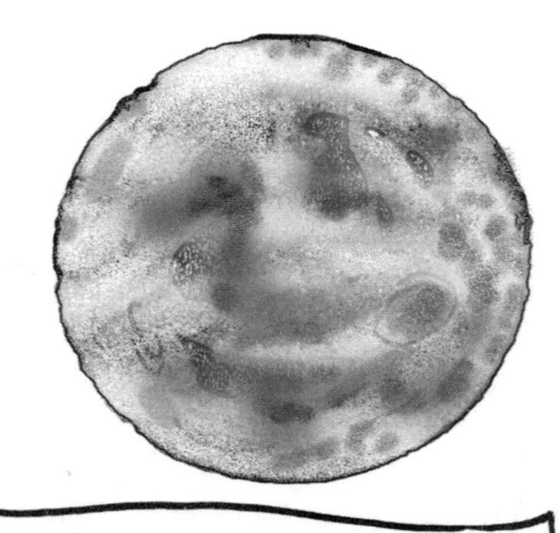

- The gravitational forces between Earth and the Moon cause some interesting effects. One of these is tides in Earth's waters. (See page 123.) Gravity also controls the rate of the Moon's rotation, and is responsible for the fact that only one side of the Moon is ever seen from Earth.
- The Moon rotates once during its revolution. Because of this, the same side always faces Earth. The other side of the moon was never seen until a Soviet spacecraft orbited the moon in 1959.

Looney Moon Questions

- Is the moon really made of green cheese?
- Who is the man in the moon? Is that his face on it?
- Why do werewolves howl at the moon?
- Can you truly cure warts by burying a dead cat by the light of a full moon?
- Is it true that the word "lunatic" comes from the Latin word for moon?

- The moon is the only extraterrestrial body visited by humans.
- Why does the moon seem to have a face? The "facial features" are areas of shadow caused by huge pits. These are called seas, though they have no water.
- The U.S. spacecraft ***Apollo 11*** landed on the Moon on July 20, 1969 on the Sea of Tranquility. Two American astronauts walked on the moon that day. Since the moon has no wind or weather, their footprints may never be disturbed.

The Moon's Phases

The Moon appears different on different nights because the positions of the Earth, Sun, and Moon change in relationship to each other as the Moon orbits Earth. These changes that we see are known as the ***phases of the Moon***. The Moon takes about $29\frac{1}{2}$ days to complete its revolution cycle.

The Moon's Phases

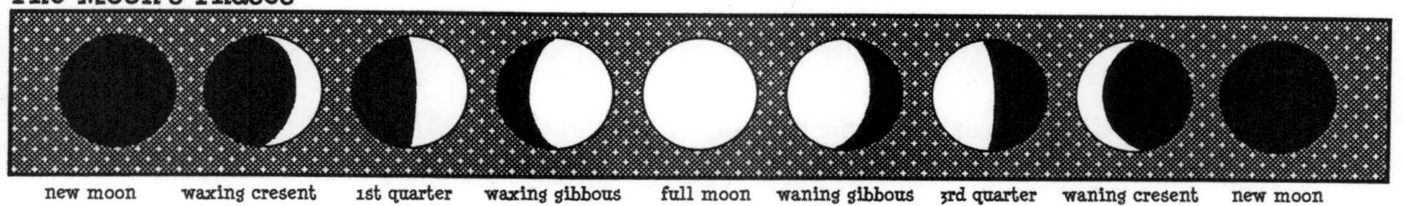

Lunar Eclipses

An ***eclipse*** occurs when the shadow of one space body falls on another body, or when one blocks the light shining on another body. In any eclipse, one space body is darkened.

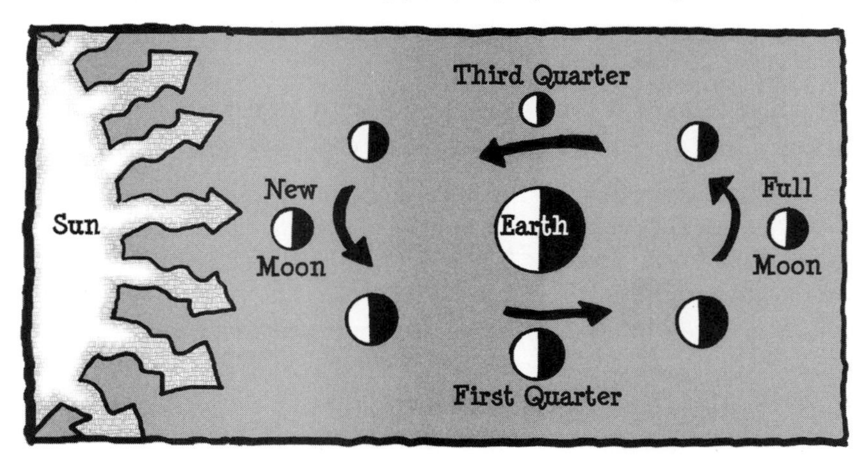

A ***lunar eclipse*** occurs when the Moon is darkened because it is passing through Earth's shadow. Earth's shadow is made up of two parts: a dark ***umbra*** and a fainter ***penumbra***. A total eclipse can be seen only when the Moon is covered by the umbra. Even during a total lunar eclipse, the Moon does not disappear completely from sight. Usually it appears to be dark red, because some of the Sun's light passes around the edges of Earth and reflects off of the Moon.

A lunar eclipse does not happen every month. As the Moon orbits Earth, it often passes below or above Earth's shadow. An eclipse occurs only when the Moon, Earth, and Sun are positioned in a straight line during a full Moon phase. A lunar eclipse can last up to an hour and forty-five minutes.

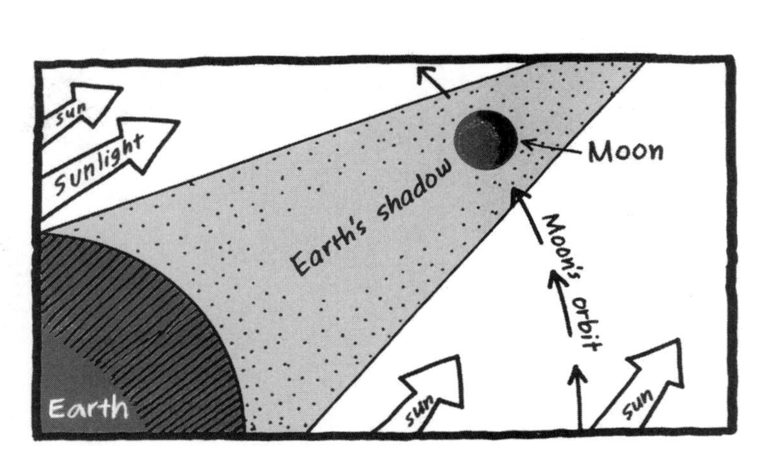

Get Sharp Tip #3

A total eclipse happens when the entire Moon passes through Earth's shadow. If only a part of the Moon passes through the shadow, the eclipse is partial.

Solar Eclipses

Imagine the shock ancient people felt when darkness appeared in the middle of the day! Imagine how frightening this must have been—before humans understood the movements of the Moon and planets. Now we know this strange occurrence as a ***solar eclipse***, and we understand why it happens. A solar eclipse occurs when the Moon's shadow falls on the Earth.

Total Eclipse

A total eclipse is seen in the area where the smaller, darker part of the shadow (the umbra) falls on Earth's surface. A total eclipse is the most unusual and spectacular kind of solar eclipse. People will travel long distances to catch a glimpse of one, because this is the only time the Sun's corona can be seen without special equipment. And the Sun's corona is quite a sight!

As the Moon's shadow moves from west to east across the Earth, the moon completely blocks out the Sun. At that moment, Sun's outer atmosphere, the ***corona***, is visible for a short time. It flashes around the darkened circle of the sun. Sun flares may be seen also during a total eclipse.

A total eclipse is brief. The average length is about two and one-half minutes. Some have lasted as long as seven minutes, but many last only a few seconds. A total eclipse can be seen only in the area of the Earth that is covered by the umbra, or the inner shadow cast by the Moon.

Partial Eclipse

A partial eclipse occurs when the Moon covers only a part of the Sun. A partial eclipse is seen in areas of Earth where only the penumbra, or fainter shadow, is cast on Earth.

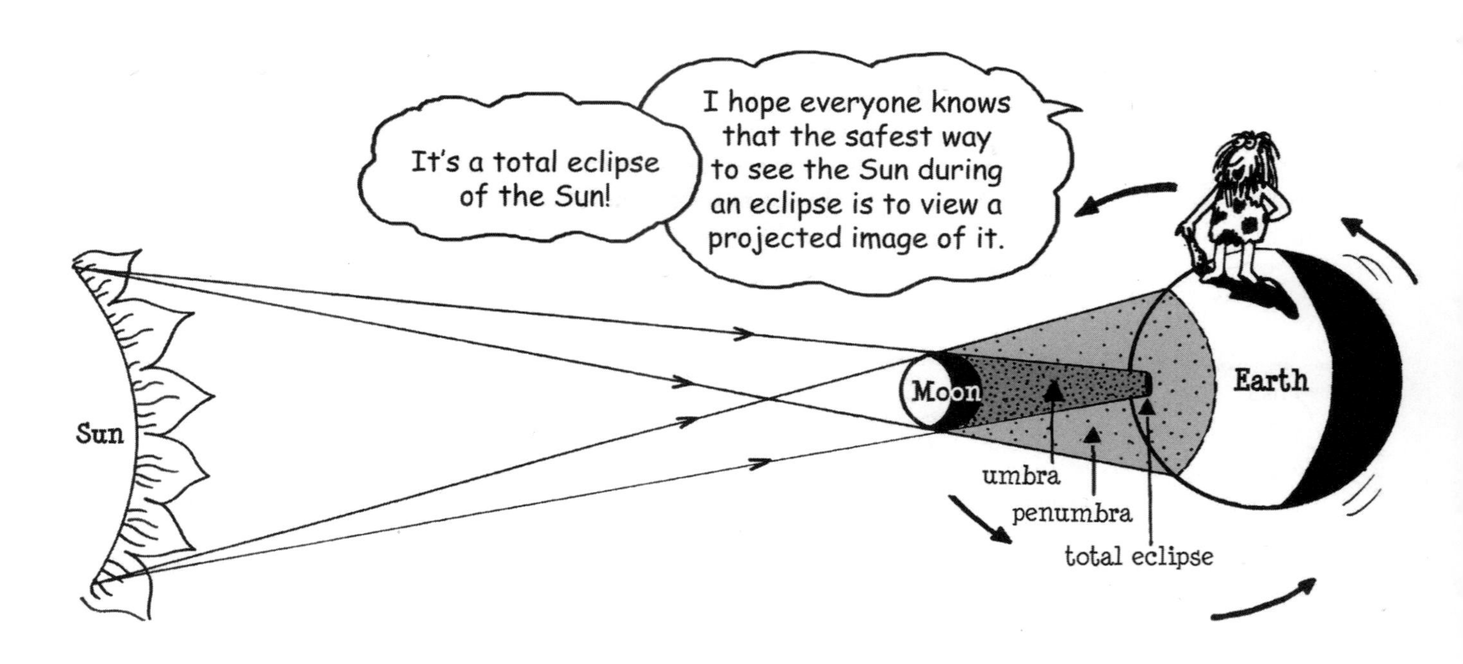

Talking Sun Sense (or How to View the Sun Safely)

1. First, choose a box about two feet long and cut a one-inch hole in the end.

2. Tape a piece of white paper to the INSIDE end opposite the hole.

3. Tape an index card over the OUTSIDE of the hole. Use a pin to punch a hole in the center.

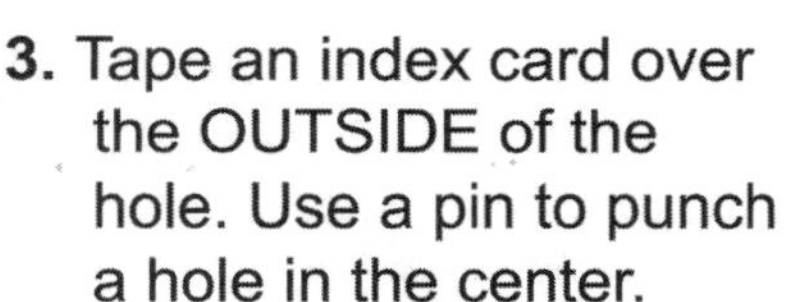

4. Adjust the box until the Sun shines through the pinhole and projects a glowing disk on the far end. Keep watching until the Moon passes across the Sun, causing the solar eclipse.

Get Sharp Tip #5

It may take a long time for the Moon to block the Sun completely. Protect yourself from the Sun's harmful rays with sunblock lotion and appropriate clothing and hats.

The Milky Way & Beyond

Our Sun is just one of billions of bright stars in space. With its planets, our Sun is part of a larger community called a galaxy. A ***galaxy*** is a system of stars, dust, and gases all held together as a group by gravity. Our own galaxy is the Milky Way. It is just one of billions of other galaxies spread throughout the universe. The smallest known galaxies may contain only 100,000 stars, but the largest known galaxy, M87, contains over three billion stars.

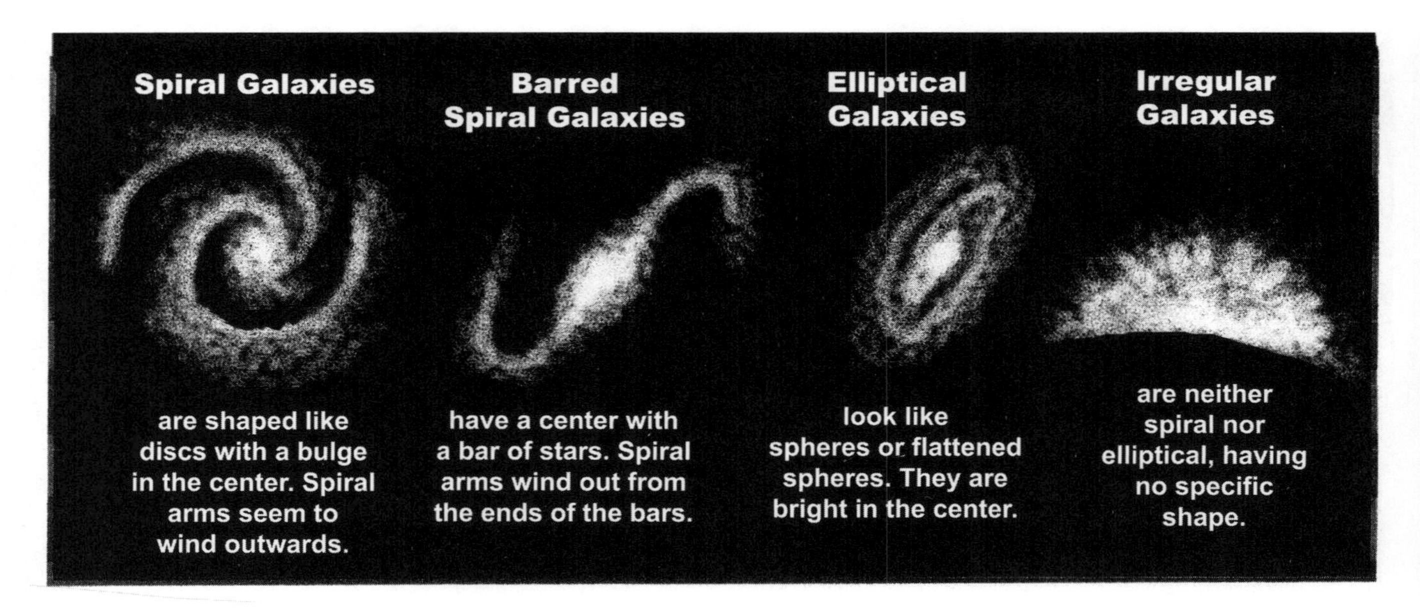

Spiral Galaxies
are shaped like discs with a bulge in the center. Spiral arms seem to wind outwards.
Barred Spiral Galaxies
have a center with a bar of stars. Spiral arms wind out from the ends of the bars.
Elliptical Galaxies
look like spheres or flattened spheres. They are bright in the center.
Irregular Galaxies
are neither spiral nor elliptical, having no specific shape.

The Milky Way is a spiral galaxy with over 200 billion stars and plenty of dust and gas in between the stars. It would take travelers like us 130,000 light years to cross the Milky Way Galaxy. (So we probably won't be seeing you anytime soon.)

The Milky Way Galaxy
(if seen from the side)

Galaxies exists in groups, called ***clusters.*** The Milky Way belongs to a cluster of 32 galaxies called the Local Group. The Milky Way is the second largest galaxy in the Local Group. Andromeda is the largest galaxy in this cluster. Clusters of galaxies seem to form larger clusters, called ***superclusters***. The Milky Way's Local Supercluster contains thousands of galaxies spread over a million light-years of distance in space.

The Universe

All these galaxies, clusters, and superclusters, when taken together, form the universe. The ***universe*** consists of everything that exists in space. It includes the Earth, the Sun, and everything in the solar system. In addition, it includes every other star and any planets it might have, all other celestial bodies (such as comets and asteroids), energy particles, magnetic fields, gases, and cosmic dust. Scientists who focus on questions about the universe are in the field of cosmology. ***Cosmology*** is the study of how the universe began, what it is like now, how it is changing, and what its future might be.

What is the Big Bang?

It is a widely held scientific theory about the origin of the universe. According to the theory, the universe began a long time ago as one dense atom. Then a big, hot explosion occurred—a series of nuclear reactions that caused that small amount of matter to blow apart. The term *Big Bang* was developed in 1950 by British scientist Fred Hoyle. Many scientists believe that after the Big Bang the matter in the universe has continued to fly apart. Will the universe fly apart forever?

What is a quasar?

Some distant galaxies have brilliant, (though mysterious) starlike objects at their cores. These objects are called *quasars*, which is short for *quasi-stellar radio sources*.

Some activity, probably due to the presence of a massive black hole at the center of a galaxy, causes this star-like object to emit enormous amounts of energy. This energy takes the form of light, radio waves, or other waves. These mysterious energy sources were first identified in 1963.

What is the Big Squeeze?

The Big Squeeze is the opposite idea from the idea that the universe will continue to expand. Other scientists wonder if the universe will become so dense that the expansion will reverse. Then, gravitational forces might begin to pull everything back together again in a Big Squeeze.

I love **cosmology**. Beginnings are always so interesting.

Stars

Stars are hot, bright, burning spheres of gas. From Earth, all stars may look like similar little spots of shimmering light. Stars may look as if they are splattered across a flat surface above us, all about the same distance from Earth. This is definitely not the case! Stars vary greatly in size, brilliance, luminosity, color, temperature, age, and distance from Earth.

Star Distances

Stars are different distances from Earth and from each other. Two stars that look as if they are next to each other may be millions of miles apart. The closest star to our solar system, Proxima Centauri, is 4.2 light years away. When you view a star in the sky, you are not seeing the star as it is now, but as it was when the light first left the star to travel across space.

Star Color

The temperature of a star affects its color. A star the size and medium-hot temperature of our Sun, for instance, would burn about 10 times hotter than a red dwarf. During the main part of their lives (main sequence), stars are classified by their surface temperatures in this way:

Blue (temperature about 60,000 K) *(Kelvins)*
White (temperature about 9,000 K)
Yellow (temperature about 6,000 K)
Red (temperature about 2,500–4,000 K)

Star Size

There is a great difference in the masses of stars from the time they are born. In star-talk, you might hear stars of different sizes referred to as ***dwarfs, medium-sized stars*** (like our Sun), ***giants,*** or ***supergiants.***

Star Luminosity

Luminosity is the rate at which a star pours out energy. It is related to the size and surface temperature of a star. Luminosity and temperature are measures used by astronomers to classify stars. Most of the known stars are less luminous than our Sun.

Star Brilliance (or Magnitude)

The brightness of a star as seen from Earth is called its ***apparent magnitude***. But since stars are not all the same distance from Earth, the appearance from Earth is not a true measure of brightness. A star's ***absolute magnitude*** is the measure of its actual brightness, determined as if it were a specific distance away (32.6 light years). Stars are given a magnitude number as a measure of brilliance. The brighter the magnitude, the lower the number will be. Magnitudes of some extremely bright stars even fall below zero.

Sirius, the brightest star in the sky, has a magnitude of -1.4. The faintest stars visible with a telescope are about Mag. 6.

The Life Story of a Star

We might think of stars as unchanging and everlasting. But they age, just like we do. Though it takes millions, billions, or trillions of years, a star has an entire life cycle—a birth, a youth, a middle age, and eventually, a death.

1 Stars are born in ***nebulae***, vast clouds of dust and gas in space. Within a nebula, blobs of dust grow and become denser. As the density and gravity increase, the temperature climbs until it ignites nuclear reactions. The reactions cause the nebula to break up into a cluster of many baby stars.

2 After the star forms, it is in its main life period, called the ***main sequence period***. A main sequence star lives and shines fairly steadily for millions of years or more.

3 Eventually, a star begins to burn itself out. It runs out of fuel and starts cooling down. As it cools, it collapses, causing temperatures to rise. The intense heat causes the gases to explode, so the star swells up into a glowing ***red giant*** that may be a hundred times the size of the original star.

4 From the red giant stage, a dwarf or medium-sized star (such as our Sun) slowly cools off and shrinks. It becomes a faint, small star called a ***white dwarf***. Eventually it will fade out altogether into a ***black dwarf***. From the red giant stage, a giant or supergiant will blow up into a huge explosion called a ***supernova***. A supernova may leave behind a tiny, dense, fast-spinning star called a ***neutron star.*** Such a star may give out radio waves in pulses as it rotates. These bursts of radiation are called ***pulsars***.

5 There is some belief that a neutron star can shrink to a body so dense that the star disappears inside itself. This is known as a ***black hole***. The idea is that the gravitational pull is so strong that everything nearby is sucked inside. Even light cannot escape.

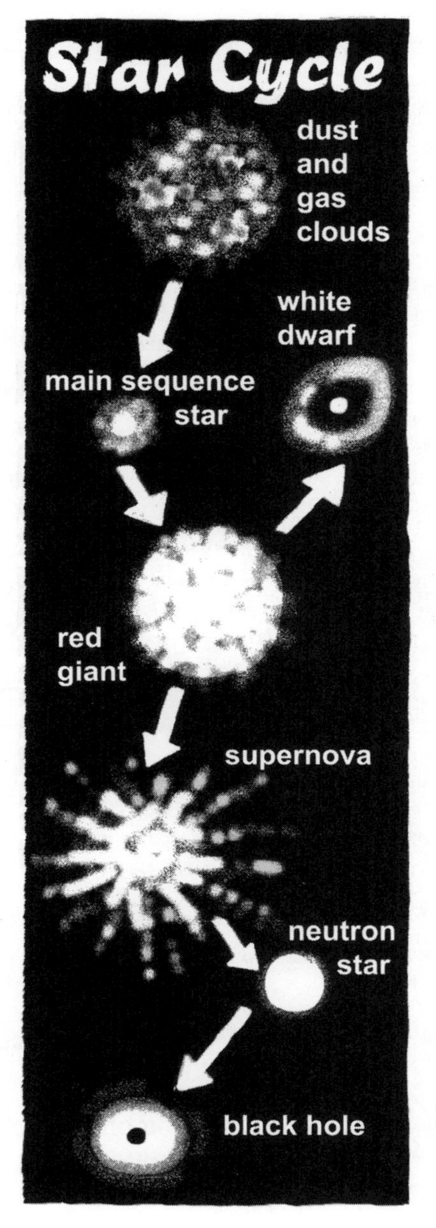

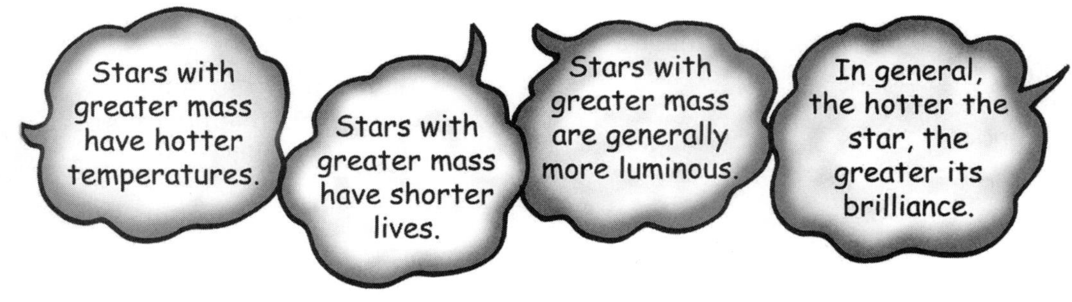

Stars with greater mass have hotter temperatures.
Stars with greater mass have shorter lives.
Stars with greater mass are generally more luminous.
In general, the hotter the star, the greater its brilliance.

Which Star is Which?

White dwarves, red giants, novas, double stars, neutron stars, quasars—so many stars! Here's a quick review to help you remember which stars are which.

protostar
A protostar is the first step in the evolution of a star. This is formed when nuclear fusion takes place in the core of the nebula, causing gas to begin glowing.

giant star
This is the label given to large stars, much larger than the Sun. Some stars are born as giants. Others become giants only near the end of their lives. (See *red giant*.)

red giant
When a star begins to cool after the main sequence period, it expands outward and puffs up into a giant star with a red glow. All stars go through a red giant phase before they die.

main sequence star
This period in a star's life begins when the mass of the new star is stabilized. This life period that lasts for millions or billions (or even trillions) of years.

blue giant
A blue giant is a star that is large and hot enough to burn at a very high temperature. A blue giant burns its fuel quickly with a very bright light. It has a short life.

Polaris, the North Star, is a variable star whose brightness changes every four days.

dwarf stars
This is the label given to the smaller stars in the universe. These stars are smaller than our Sun.

supergiant
Some stars are even bigger than giants, and are called supergiants. A massive star may become a supergiant as it gets older and begins to burn helium. This fuel burns hotter, so the star puffs out farther than normal.

red dwarves
The most common type of star in the universe is the red dwarf. Their small size results in cool temperatures, which leads to the red color. Most of them don't look red because they are too far away for the color to show.

nebula
A nebula is a low-density cloud of gas and dust from which stars are born.

medium stars
(also known as yellow stars)
Medium stars are similar in size to our Sun. They burn yellow because they have a medium temperature.

white dwarf
The dense white dwarf stage occurs after the red giant phase in small and medium stars. The star's matter collapses inward due to the pull of gravity, and the star becomes extremely dense. It shines for a while with a dim white light.

supernova
A supernova is a powerful explosion of a star that lights up the sky with tremendous heat and light. A supernova results from the collapse of a star that has massive size (larger than our Sun).

black hole
This is an extremely dense object that forms when a massive star collapses inward due to an immense gravitational force. When a giant star dies, it explodes in a supernova. What is left of the star gets very small, with so much gravitational pull that all the star's matter is pulled inside itself.

quasar
There are some extremely bright star-like objects at the very edges of the universe. These are objects about the size of our solar system that shine brighter than most galaxies because they produce so much light and energy.

variable star
Some stars change, or appear to change, in their brightness. This may be because the star expands and contracts. It may also occur because two or more stars orbiting a common point eclipse each other and the light is blocked out temporarily.

black dwarf
This is the phase when a star has totally burned out so that it no longer shines. It is the final phase in the life of a small or medium star.

neutron star
The core of a massive star ends up as a neutron star after a supernova. A neutron star is a rapidly-spinning star that gives off radio waves.

pulsar
Some neutron stars give off radio waves in a pulsing rhythm. These are called pulsars.

binary stars
This is a pair of stars that orbit each other or appear to orbit each other.

When in Doubt— Talk it Out!

Did you ever wonder what would happen if we got too close to a **black hole?**
It wouldn't be pretty! First a tremendous force would suck us in. Then, we'd get stretched out like a string of spaghetti. You and I, and everything else, would never get out! Even light couldn't escape.
If we're smart, we'll stay far away from black holes.

Patterns in the Sky

Long ago, early stargazers invented ***constellations***. They looked into the skies and imagined shapes or patterns created by groups of stars. Names of animals, ancient gods, people, or other creatures were given to these groups of stars called constellations. These names, plus some new ones, are still used today by star watchers.

Each constellation has a Latin name, but most are better known by a common English name. The sky contains 88 different constellations. The largest one is Hydra (the Water Snake) and the smallest one is Crux (the Cross). Often, a constellation is recognized by the brightest star or stars within it. In the constellations shown here, the brightest stars are named.

The Northern Sky in January

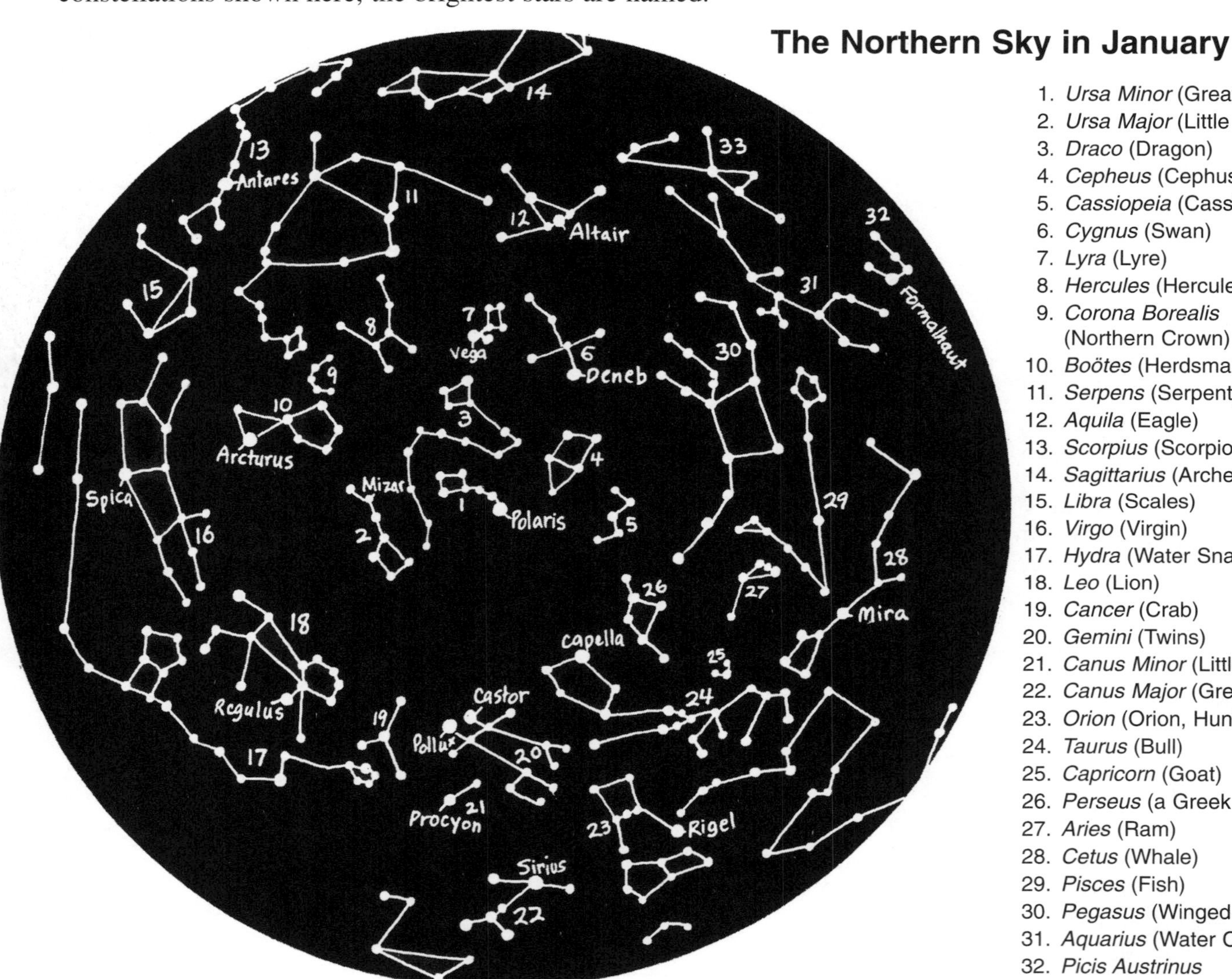

1. *Ursa Minor* (Great Bear)
2. *Ursa Major* (Little Bear)
3. *Draco* (Dragon)
4. *Cepheus* (Cephus)
5. *Cassiopeia* (Cassiopeia)
6. *Cygnus* (Swan)
7. *Lyra* (Lyre)
8. *Hercules* (Hercules)
9. *Corona Borealis* (Northern Crown)
10. *Boötes* (Herdsman)
11. *Serpens* (Serpent)
12. *Aquila* (Eagle)
13. *Scorpius* (Scorpion)
14. *Sagittarius* (Archer)
15. *Libra* (Scales)
16. *Virgo* (Virgin)
17. *Hydra* (Water Snake)
18. *Leo* (Lion)
19. *Cancer* (Crab)
20. *Gemini* (Twins)
21. *Canus Minor* (Little Dog)
22. *Canus Major* (Great Dog)
23. *Orion* (Orion, Hunter)
24. *Taurus* (Bull)
25. *Capricorn* (Goat)
26. *Perseus* (a Greek Hero)
27. *Aries* (Ram)
28. *Cetus* (Whale)
29. *Pisces* (Fish)
30. *Pegasus* (Winged Horse)
31. *Aquarius* (Water Carrier)
32. *Picis Austrinus* (Southern Fish)

Magnitudes of Brightest Stars

(The lower the number, the brighter the star)

Star	Magnitude	Star	Magnitude	Star	Magnitude
Aldebaron	= 1.1	Deneb	= 1.3	Procylon	= 1.2
Altair	= 0.9	Formalhaut	= 1.3	Regulus	= 1.3
Antares	= 1.2	Mira	= variable	Rigel	= 0.3
Arcturus	= 0.2	Mizar	= 2.2	Sirius	= –1.6
Capella	= 0.2	Polaris	= 2	Spica	= 1.2
Castor	= 2	Pollux	= 1.2	Vega	= 0.1

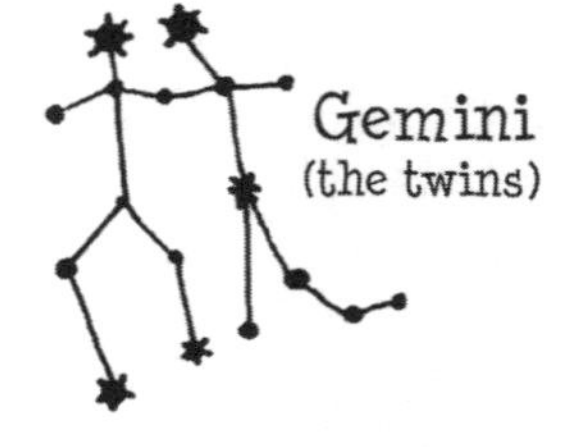

The stars in the sky appear to be moving constantly. This is because the Earth is spinning. These maps of the stars will give you an idea of the locations and shape of some of the constellations.

Not all the constellations are visible at the same time or from the same location every night. Some constellations are seen only in the Southern Hemisphere or Northern Hemisphere. The stars will appear at different places in different seasons.

The Southern Sky in January

1. *Crux* (Cross)
2. *Centaurus* (Centaur)
3. *Lepus* (Hare)
4. *Ara* (Altar)
5. *Triangulum Australe* (Southern Triangle)
6. *Octans* (Octant)
7. *Libra* (Scales)
8. *Scorpio* (Scorpion)
9. *Sagittarius* (Archer)
10. *Ophiuchus* (Serpent Holder)
11. *Hercules* (Hercules)
12. *Corona Australis* (Southern Crown)
13. *Boötes* (Herdsman)
14. *Virgo* (Virgin)
15. *Hydra* (Water Snake)
16. *Leo* (Lion)
17. *Cancer* (Crab)
18. *Carina* (Keel)
19. *Canis Major* (Great Dog)
20. *Canis Minor* (Little Dog)
21. *Gemini* (Twins)
22. *Orion* (Hunter)
23. *Taurus* (Bull)
24. *Eridanus* (River Eridanus)
25. *Lupus* (Wolf)
26. *Aries* (Ram)
27. *Cetus* (Whale)
28. *Pisces* (Fish)
29. *Phoenix* (Phoenix)
30. *Grus* (Goat)
31. *Aquarius* (Water Carrier)
32. *Pegasus* (Winged Horse)
33. *Aquila* (Eagle)
34. *Delphinus* (Dolphin)
35. *Cygnus* (Swan)

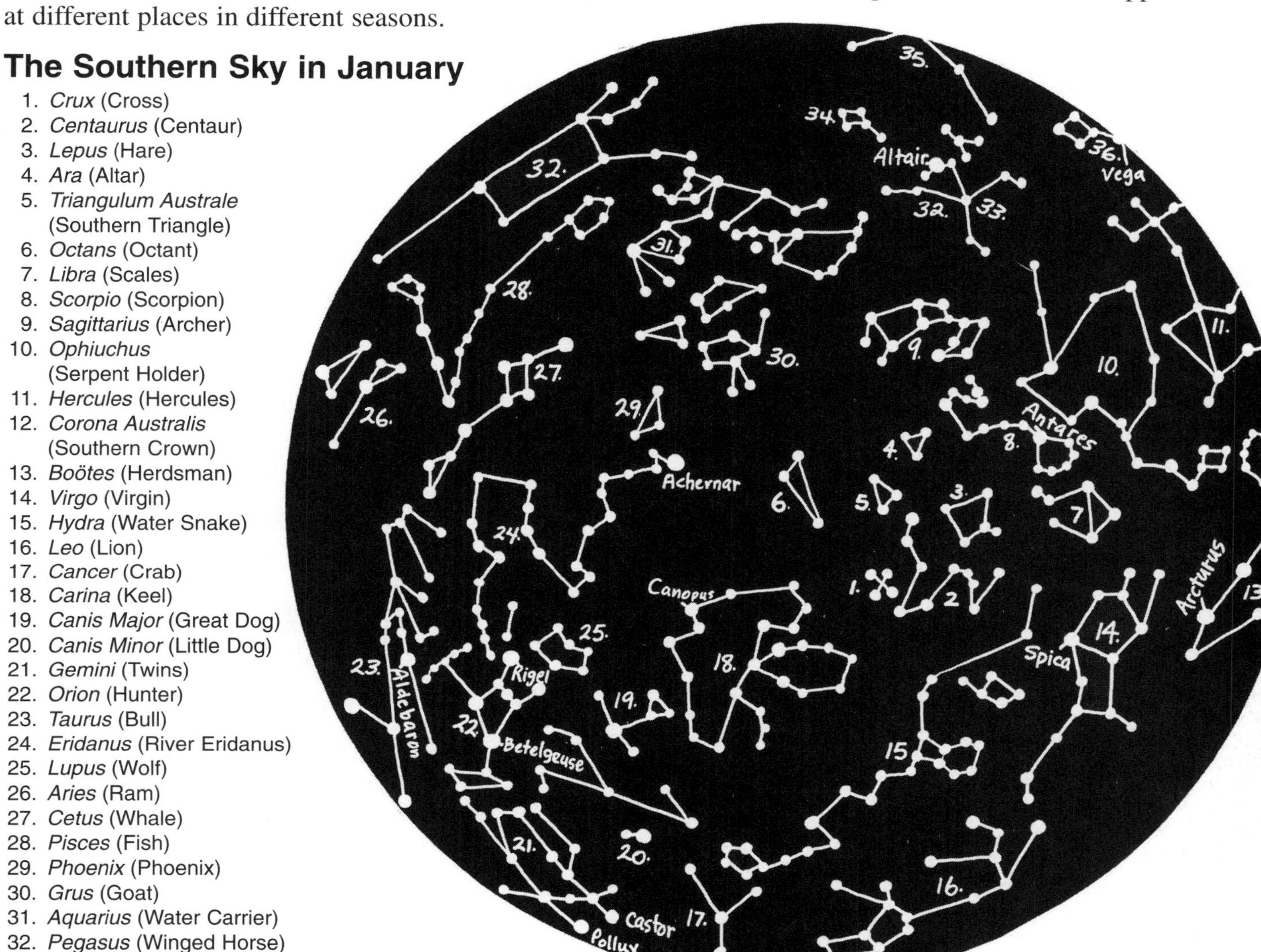

Magnitudes of Brightest Stars

(The lower the number, the brighter the star)

Achnernar = 0.6
Aldebaron = 1.1
Altair = 0.9
Antares = 1.2
Arcturus = 0.2
Betelgeuse = 0.9
Canopus = –0.9
Castor = 2
Pollux = 1.2
Rigel = 0.3
Spica = 1.2
Vega = 0.1

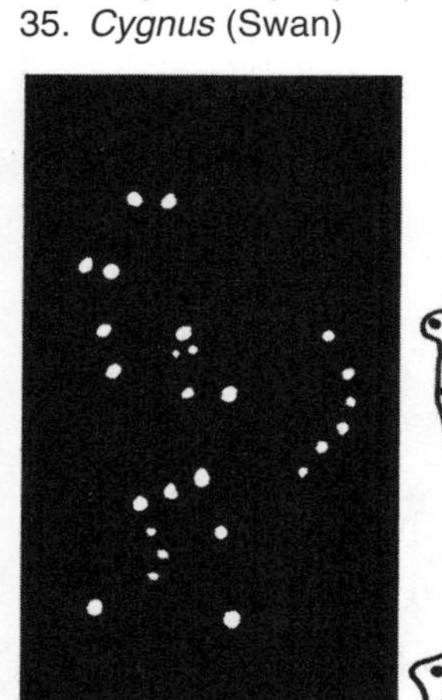

Orion (the hunter)

Leo (the lion)

Exploring the Skies

As you can imagine, the skies have been a source of wonder and mystery for people from the beginning of life on Earth. Human curiosity has turned eyes toward space for centuries.

A Timeline of Some Major Events & Discoveries

B.C.

3000 The Babylonians kept astronomical records.

400 Aristotle showed that the Earth is round.

300s Aristotle developed a system of astronomy.

240 The comet that later became known as Halley's Comet was first sighted.

A.D.

140 Ptolemy put forth his theory that the Earth was the center of the universe.

1543 Copernicus claimed that the Sun, not the Earth, was at the center of the universe.

1600s Kepler showed that planets move in elliptical orbits.

1608 The first telescope was made by Dutchman Hans Lippershey.

1609 Galileo examined the sky with a telescope.

1671 The reflecting telescope was invented by Newton.

1675 The Royal Greenwich Observatory was founded in England.

1781 Uranus was discovered. (It was the first planet discovered with a telescope.)

1845 German astronomer Johann G. Galle discovered the planet Neptune.

1914 Eddington identified spiral galaxies for the first time.

1920s Edwin Hubble demonstrated that the universe is expanding.

1926 American Robert Goddard launched the first liquid-propellant rocket.

1930s Physicist Hans Bethe explained how nuclear fusion powers stars.

1930 American astronomer Clyde W. Tombaugh discovered the planet Pluto.

1957 The U.S.S.R. launched *Sputnik 1*, the first artificial satellite.

1957
A Space Race First

A dog named Laika was the first living creature to travel into space. She was launched by the Soviet Union aboard ***Sputnik 2*** on November 3, 1957. The mission did not include a way to return the dog to Earth. So Laika died after a few days.

1958
Monkeying Around In Space

A monkey named Gordo was launched into space aboard a ***Jupiter*** flight. Scientists wanted to explore the safety of space flights for living creatures. Gordo suffered no adverse effects from the space trip, but a failure in the rocket prevented his recovery.

I'm clipping articles for my scrapbook.

1959
Animals Survive Space Trip

Two space monkeys, Able and Baker, made history as the first living beings to be successfully recovered after a space flight. They were launched on a ***Jupiter*** mission and traveled 1,500 miles in space.

1961
CHIMP MAKES HISTORY

A chimpanzee named Ham made an 18-minute flight in a ***Mercury*** space capsule launched on January 31, 1961. Ham ended the mission in good shape.

1973
Zoo Blasts Off

Two spiders, Anita and Arabella, took a trip aboard the ***Space Lab.*** They were not alone! 720 fruit flies, 6 mice, 2 minnows, and many minnow eggs accompanied them. Other trips have taken bullfrogs, hamsters, rats, dogs, cats, monkeys, and chimps.

1957 The U.S.S.R. launched *Sputnik 2*, the first satellite carrying a living being (a dog) into space.

1958 NASA *(The National Aeronautics and Space Administration)* was formed.

1958 *Explorer 1* was the first satellite launched by the United States.

1959 The U.S.S.R. launched *Luna 2*, the first space probe to orbit the Moon.

1961 Soviet cosmonaut Yuri A. Gagarin was the first person to orbit the Earth.

1961 Alan B. Shepard, Jr., was the first American launched into space.

1962 John H. Glenn, Jr., was the first American to orbit Earth.

1962 *Ranger 4* was the first U.S. space probe to hit the Moon.

1962 *Telstar 1*, the first communications satellite, was launched.

1963 Soviet cosmonaut Valentina Tereshkova became the first woman in space.

1963 Dutch astronomer Maarten Schmidt identified *quasars.*

1965 Alexsei Leonov, traveling in *Voshkod 2*, performed the first space walk.

1967 Three *Apollo* astronauts were killed in a fire that occurred on the launch pad before their rocket launched.

1967 British astronomer Jocelyn Bell Burnell identified *pulsars.*

1968 The U.S. launched *Apollo 8,* the first manned spacecraft to orbit the Moon.

1969 On July 20, U.S. astronauts Neil A. Armstrong and Buzz Aldrin became the first humans to land on the Moon.

1971 The NASA space probe, *Mariner 9*, was the first spacecraft to orbit Mars.

1971 The Soviet space probe, *Mars 3*, was the first craft to land on Mars.

1975 Soviet and U.S. spacecraft linked in space as part of the first international space mission.

1977 U.S. probe *Voyager 2*, was launched. Over the next several years, this probe flew past and photographed Jupiter, Saturn, Uranus, and Neptune.

1981 U.S. launched the space shuttle *Columbia*, the first reusable piloted spacecraft.

1983 U.S. launched *Pioneer 10,* which traveled beyond all the planets.

Finally, an article with a frog in it!

Tourists in Space: A Good Idea?

Dennis Tito was the first space tourist. But will he be the last? There is increased interest in space travel among adventurers who have enough money to buy their way into outer space. Is this a blossoming new industry? Should tourists go into space? What are the dangers and advantages of this practice?

Learn a lot more about the solar system and outer space from the NASA websites. The information is fun and fascinating!

www.nasa.gov

www.nasakids.gov

And don't pass up a chance to learn about the amazing discoveries of the Hubble Space Telescope. Visit the Hubble site.

www.hubblesite.com

1985 U.S. space shuttle *Challenger* exploded after launch, killing all seven crew members.

1986 The first module of the *Mir Space Station* was launched.

1988 NASA resumed the U.S. space program with the launch of the shuttle, *Discovery.*

1989 The U.S. launched the probe *Galileo*, which reached Jupiter in 1995.

1990 U.S. space probe *Magellan* began its orbit of Venus.

1990 The *Hubble Space Telescope* was launched.

1990–1994 U.S. space probe *Magellan* mapped the surface of Venus.

1994 Comet Shoemaker-Levy collided with Jupiter.

1996 *Mars Pathfinder* sent important data about the planet back to Earth.

1997 The U.S. launched the probe *Cassini*, due to reach Saturn in 2004.

1998 The first part of the *International Space Station* was launched into place.

1999 The *Mars Polar Lander/Deep Space 2* spacecraft, which was to set down near Mars' south polar cap, was lost during landing.

2001 The first full-time crew (one American astronaut and one Russian cosmonaut) occupied the *International Space Station*.

2001 Space tourist Dennis Tito paid for and took a trip to the *International Space Station.*

2001 The *Mir Space Station* was destroyed (purposely).

2001 *Quaoar*, a large celestial body beyond Pluto, was discovered. It is the largest body found since the 1930 identification of Pluto.

2002 Scientists discovered 11 previously unseen moons orbiting Jupiter.

2003 A satellite captured the first pictures of a gamma ray burst.

2004 Sedna, the most distant object ever seen in the solar system, was seen beyond Pluto.

2004 Scientists calculated the age of the galaxy more accurately—13.6 billion years old.

2004 NASA landed two rovers on the surface of Mars.

GET SHARP

in

EARTH SCIENCE

The Earth's Structure

Viewed from space, Earth looks like a sphere covered with clouds and water. With a closer look, an observer can see large masses of land surrounded by water.

Basically, Earth is a huge ball of rock, covered with water and land, and wrapped in an envelope of air.

Layers

One way to describe Earth's structure is to tell about each of its layers.

The **crust** is a relatively thin layer of rock, ranging from 3–31 miles thick. The crust is thinnest beneath the oceans. (This is oceanic crust.) Crust beneath the continents (continental crust) is thicker. Earth's crust is thickest beneath mountain ranges.

The **mantle** is a layer of hot rock (magma) about 1,800 miles thick. The uppermost part is rigid. The lower mantle region is a material with the consistency of a very thick liquid.

The **core** is the center region of Earth, about 2,200 miles thick. It is extremely hot rock that is mostly iron and nickel. The innermost part of the core is solid; the outer core is liquid.

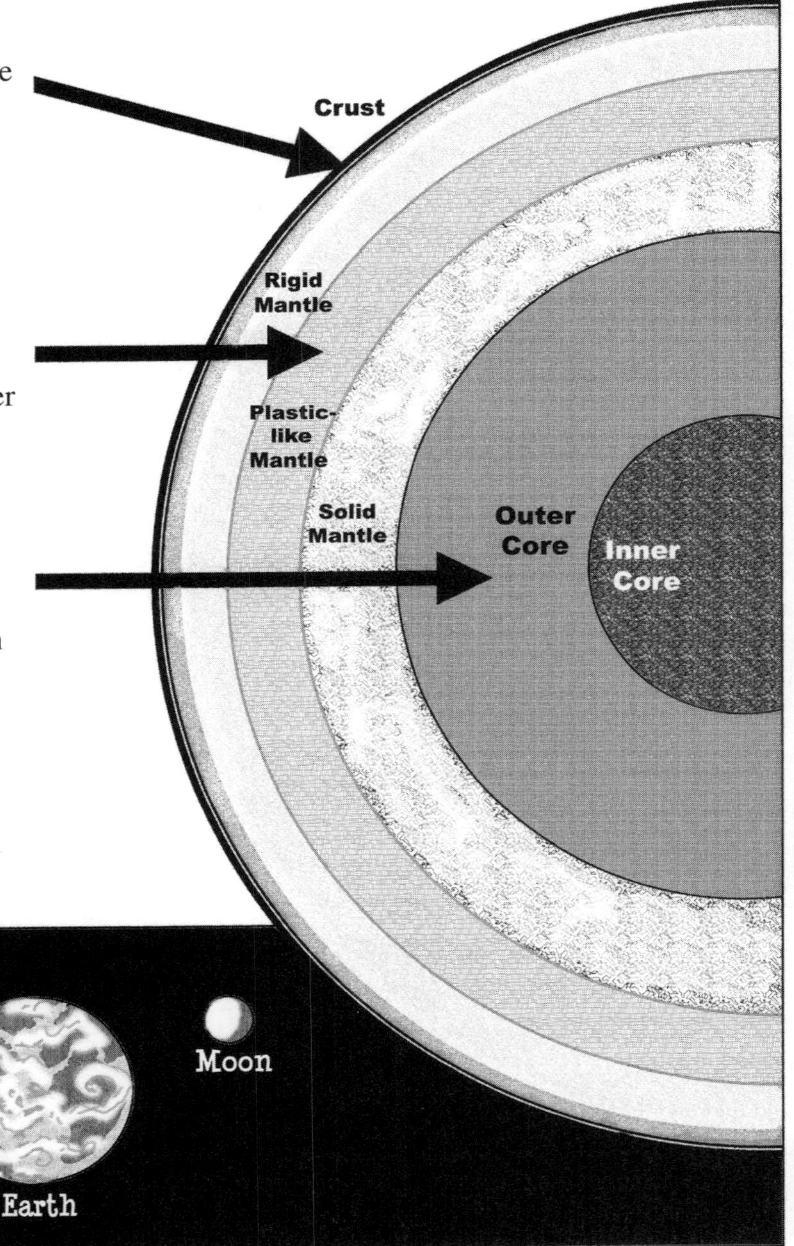

Biosphere is a word used to describe the regions of Earth's land, water, and atmosphere that are inhabited by living things.

Earth's Vital Statistics

Age — about 4.5 billion years

Mass — 6.6 sextillion tons

Polar circumference — 24,850 miles

Equatorial circumference — 24,902 miles

Distance to the center — about 4,000 miles

Total surface area — 196,900,000 miles

Land Area — 57,100,000 miles

Water Area — 139,800,000 miles

Highest land point — Mt. Everest, 29,035 feet

Lowest land point — Dead Sea, 1,310 feet below sea level

Core temperatures — 5000° to 9000° F

Surface temperature — –129° to 136° F

Lowest point — Mariana Trench, Pacific Ocean, 25,840 feet below sea level

Soil

Most land surfaces are covered with soil. Soil may be just a thin layer on top of rock, or it may be over 100 feet thick. ***Soil*** is a mixture of decayed organic material, weathered rock, air, and water. Soil is formed when wind, ice, or running water break down rocks and other materials. This process takes hundreds of years. Different soil types have different colors, textures, and chemical makeups. Soil can be waterlogged or well-drained. It can be rich or poor in organic matter.

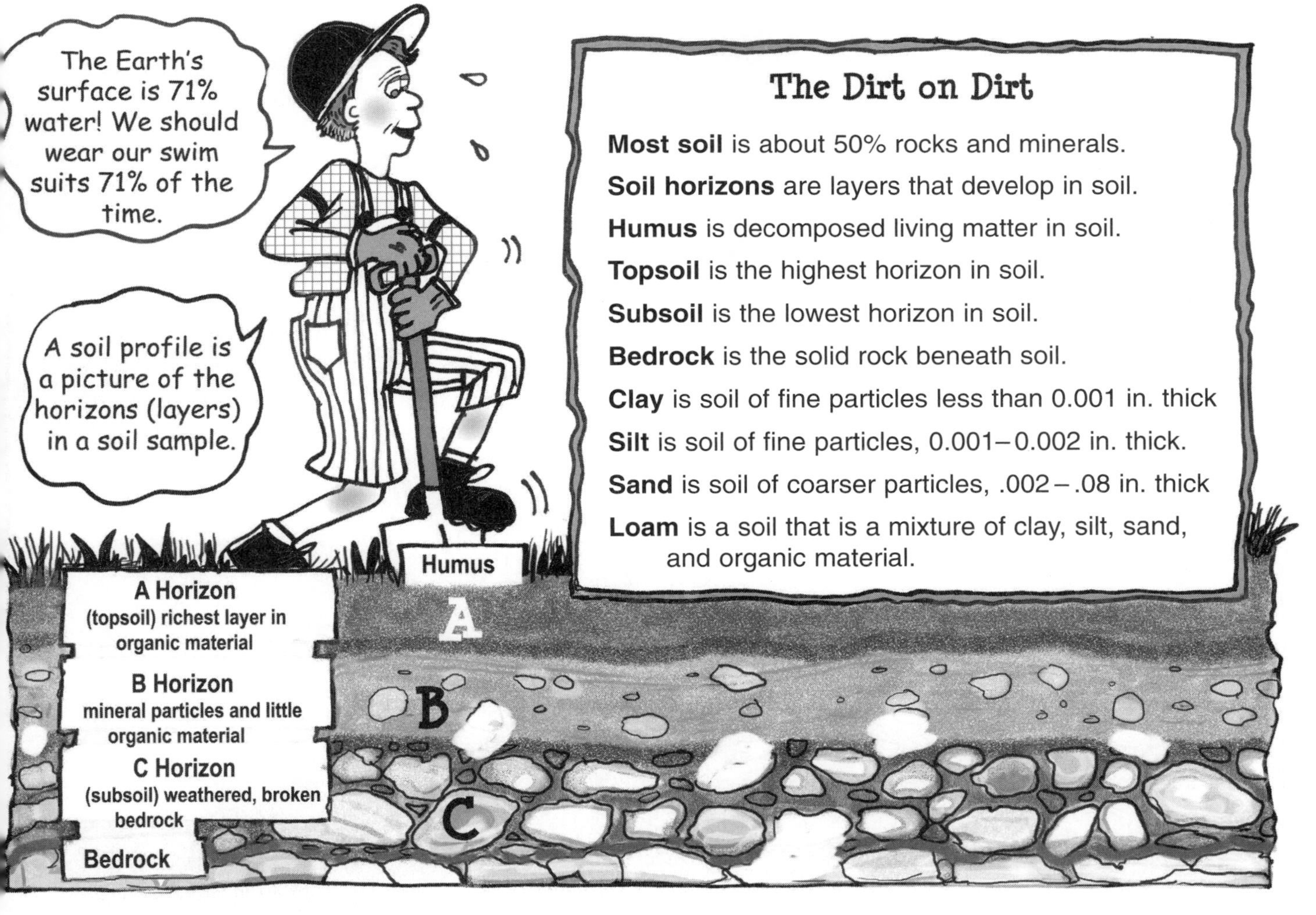

The Earth's surface is 71% water! We should wear our swim suits 71% of the time.
A soil profile is a picture of the horizons (layers) in a soil sample.
Humus
A
B
C
A Horizon (topsoil) richest layer in organic material
B Horizon mineral particles and little organic material
C Horizon (subsoil) weathered, broken bedrock
Bedrock

The Dirt on Dirt

Most soil is about 50% rocks and minerals.

Soil horizons are layers that develop in soil.

Humus is decomposed living matter in soil.

Topsoil is the highest horizon in soil.

Subsoil is the lowest horizon in soil.

Bedrock is the solid rock beneath soil.

Clay is soil of fine particles less than 0.001 in. thick

Silt is soil of fine particles, 0.001–0.002 in. thick.

Sand is soil of coarser particles, .002–.08 in. thick

Loam is a soil that is a mixture of clay, silt, sand, and organic material.

Internal Processes

> **Get Sharp Tip #5**
> The word tectonics comes from a Greek word that means builder.

The Earth is always changing. Many of the changes begin deep inside the Earth. One theory explains many of these changes.

Plate Tectonics

The ***Theory of Plate Tectonics*** is a scientific explanation for the movements within Earth. It explains processes such as earthquakes, volcanoes, mountain building, and continental drift. This theory suggests that Earth's crust, like a giant jigsaw puzzle, is made up of huge rigid plates that lie or "float" on the plastic-like lower layer of the mantle. It is believed that these plates move constantly, usually at a slow pace. They slide past each other, slide under or over one another, push together, scrape against each other, and pull apart. The energy created by movement of the plates causes many of the changes or activity we see on Earth's surface.

Internal Forces

Compression –	Tension –	Shearing –
forces that push against a solid substance directly and evenly from opposite sides and squeeze it until it folds.	force that pulls a solid apart or stretches it.	forces that are not directly opposite pushing against a solid from different sides, resulting in tearing and twisting.

Most of the activity beneath the Earth's surface (such as volcanoes or earthquakes) happens at the edges, or boundaries, of the plates.

Shake, rattle, and roll at the **No Boundaries Cafe.**

Convergent Boundaries are boundaries between two colliding plates. When two oceanic plates collide (or one oceanic and one continental), one is forced down. The area where the plate submerges is the ***subduction zone.*** When two continental plates collide, the rocks crumble and rise, forming folded mountains.
Divergent boundaries are boundaries where two plates pull apart.
Transform faults are boundaries where plates slide or scrape past each other. These are often areas of earthquake activity!
I think I'll order the blue plate special.
P
S

Continental Drift

There is a belief that the continents were once connected in a ***supercontinent*** sometimes called ***Pangaea***. The ***Continental Drift Theory*** suggests that the continents moved apart (leaving ocean basins between them), and that they are still moving.

Get Sharp Tip #6

Orogeny is the name given to the processes that build mountains. These include volcanic activity, folding, crumpling, and unwarping.

Seafloor Spreading

On the Atlantic Ocean, the ridge of the Mid-Atlantic Mountain Range is on the boundary of two plates. As the plates move away from a crack in the crust, melted magma flows into the crack and hardens to form new seafloor. This process is called ***seafloor spreading***.

Continental Drift Theory

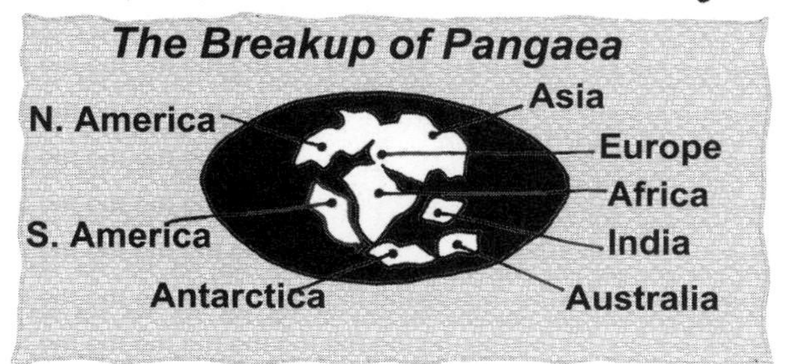

Mountain Building

Mountains are formed when molten rock from inside the Earth pushes a dome in the surface, or when it pours out of the surface to form a cone. Mountains are also formed when internal forces thrust some of Earth's crust upwards or cause the crust to fold.

Earthquakes & Volcanoes (See pages 102–105.)

Fractures are breaks in rock that result from some sort of force beneath Earth's surface.

Folds are bends in the rock. When rock is more elastic, it may bend into folds that alternate ridges (***anticlines***) with troughs (***synclines***).

Fractures, Folds, & Faults

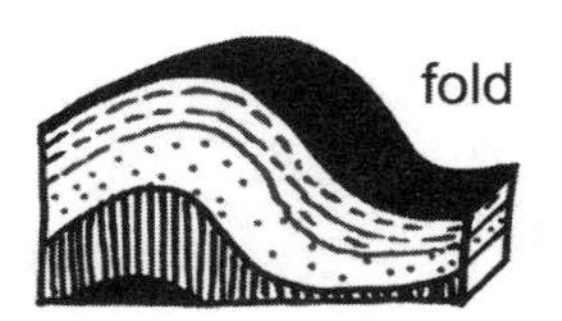

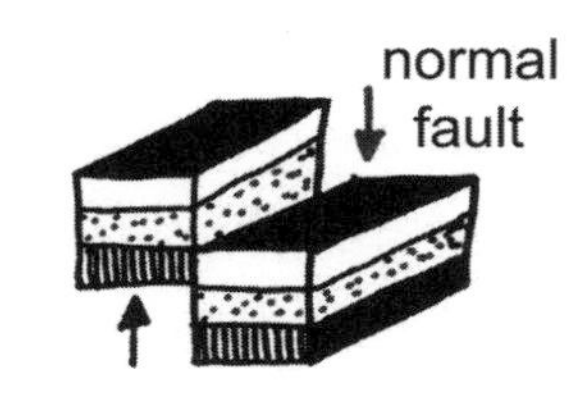

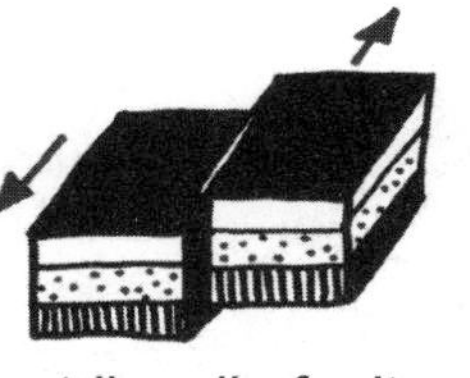

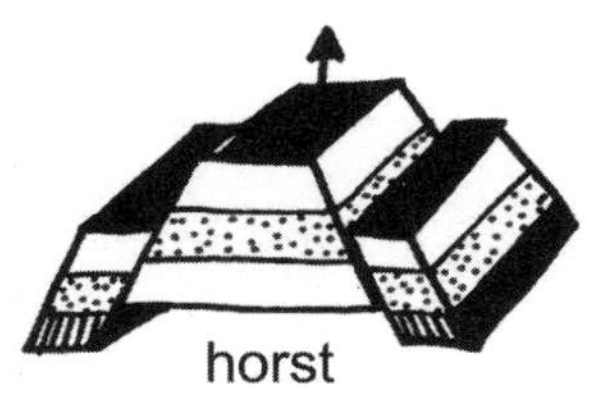

Faults are fractures along which some movement takes place. Movement can occur in any direction along the fault. Faults occur in areas where Earth's rocky crust is weak.

Earthquakes

Earthquakes are among the most frightening and damaging of the natural processes. An earthquake occurs when rocks move on opposite sides of a ***fault***, or break, in the crust. The actual quake is the shaking caused by the vibrations resulting from the break.

How an Earthquake Travels

When rock breaks within Earth's lithosphere, it releases energy in the form of waves that travel out from the fault in all directions. These are called ***seismic waves***.

The Primary Waves (**P** *waves,* or *compressional waves*) are underground. These travel fast through the earth. They move back and forth in the direction the wave is traveling.

Secondary waves (**S** *waves,* or *shear waves*) are also underground waves. These move in a side-to-side motion, traveling more slowly than the primary waves.

Surface waves travel along the surface in a rolling motion.

Get Sharp Tip #7

The **hypocenter** is the center of the earthquake deep within the Earth where the shock waves originate. The **epicenter** is the point on Earth's surface above the hypocenter. The most violent shaking is felt near the epicenter.

As many as 8,000 earthquakes might occur each day. Most of those are not strong enough to be felt.

The strongest earthquake on record was in Chile, May 22, 1960. Its moment magnitude was recorded at 9.5. It reached a magnitude of 8.3 on the Richter scale.

Studying & Measuring Earthquakes

Scientists who study earthquakes are called ***seismologists***. They use machines to determine the size of a quake. The ***magnitude*** is a measure of the energy released by the earthquake at its source. The magnitude of a quake will be the same no matter where a person is in relation to the epicenter. The ***intensity*** of a quake is a measure of the amount of shaking and damage. This measure differs by location.

The Richter Scale is commonly used to compare the sizes of earthquakes. This is a mathematical formula that calculates magnitude by recording the amplitude of seismic waves during a certain period of time. A machine called a ***seismograph*** detects earthquakes and measures their magnitudes. It records a zig zag line that shows the strength of the ground movements. The ***Moment Magnitude Scale*** is another way to find earthquake size. It measures large earthquakes more precisely than the Richter Scale because it measures more of the ground movements.

Earthquake Damage

Earthquakes can lead to tremendous loss of life and destruction to property. An earthquake can cause damage in many ways. The ground may crack. Buildings, bridges, dams, and other structures may collapse. Broken glass, falling objects, and landslides can cause injury or damage. Huge ocean waves (tsunamis) created by the vibrations can cause floods. Fallen power lines and cracked gas pipes can cause fires. Hazardous chemicals may be spilled during earthquakes or sewage may seep into water supplies from broken sewer lines.

Aftershocks

Small earthquakes often follow a large earthquake. These "follow-up" tremors are called ***aftershocks***. There are usually several of these following a major quake. Sometimes the aftershocks can be almost as strong and can cause as much damage as the original quake.

Earthquake!

What to do in an earthquake . . .

- **DROP, COVER, AND HOLD ON!**
- Crawl to a safe place under a strong table or in a doorway away from the outside of the house and away from windows or heavy objects that can fall on you.
- Stay indoors until the shaking stops and you're sure it's safe to exit.
- Stay away from windows.
- If you are outdoors, drop to the ground. Try to crawl or scurry to a clear spot away from power lines, buildings, and trees.
- When the shaking stops, take care of any injuries or call for help if injuries are serious.
- Extinguish small fires, and turn off any gas.
- Be ready for aftershocks. When you feel one, **DROP, COVER, AND HOLD ON.**

According to the website guinnessworldrecords.com, the most devastating tsunami in modern history struck Southeast Asia on December 26, 2004.
The fires that erupted after the San Francisco Earthquake of 1906 took many lives and destroyed a great portion of the city. It is believed that more people died from the fires after the quake than from the earthquake itself.
The deadliest earthquake in modern times was in Tang-shan, China in 1976. About 250,000 people were killed.
Uh, oh...
For more fascinating information about earthquakes, visit the U.S. Geological Survey earthquake site for kids: http://earthquake.usgs.gov/4kids/

Volcanoes

A ***volcano*** is another one of those sudden, dramatic, violent, and dangerous processes of nature. A volcano is any natural opening in Earth's crust through which hot melted rock (magma), fragments of rock, ash, and hot gases erupt and pour out over the surrounding land.

Volcanoes can be very destructive to life and land. Volcanoes have killed thousands of people and buried entire villages. A volcanic eruption can trigger a tsunami, causing even more death and devastation.

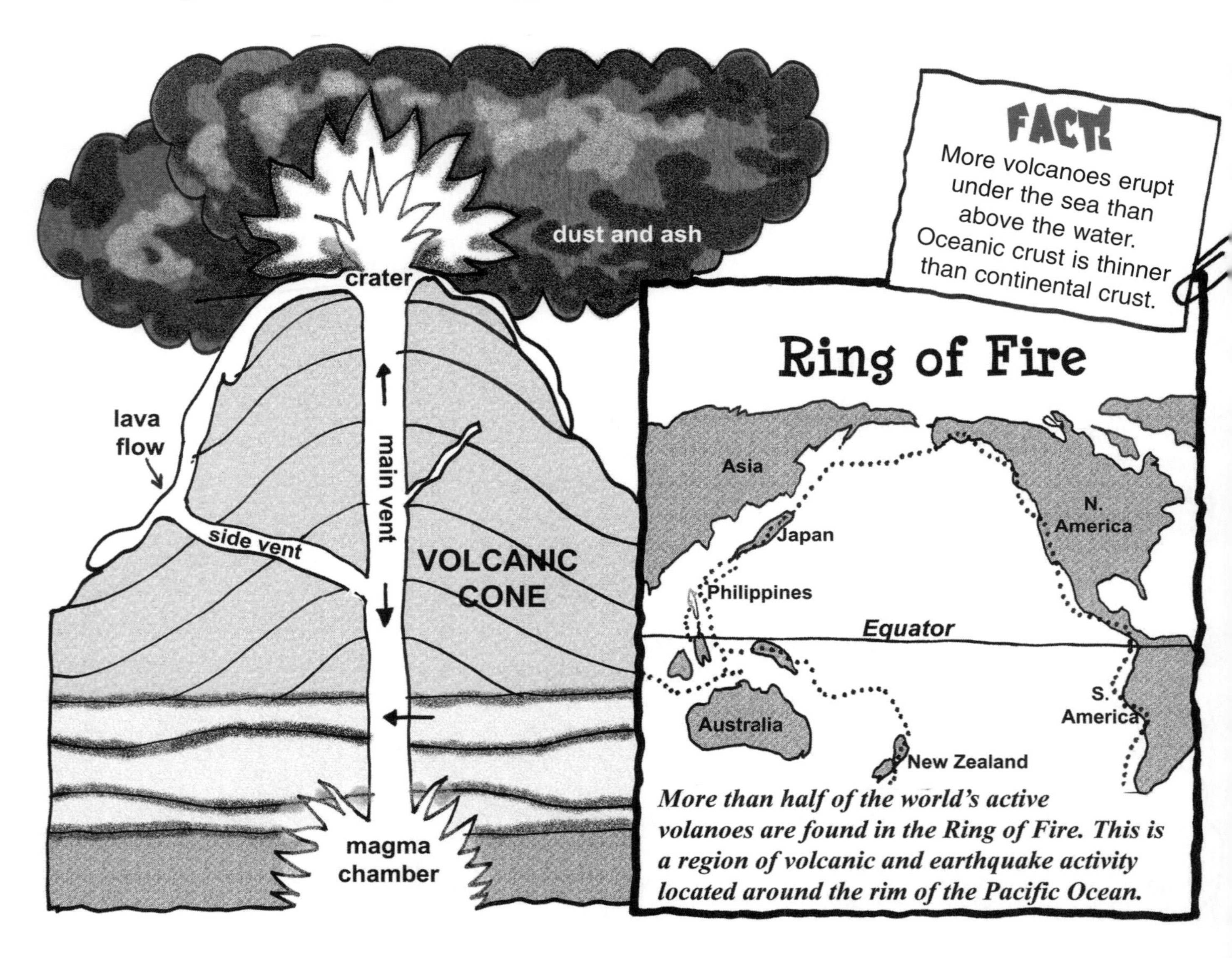

dust and ash
crater
lava flow
main vent
side vent
VOLCANIC CONE
magma chamber
FACT!
More volcanoes erupt under the sea than above the water. Oceanic crust is thinner than continental crust.
Ring of Fire
Asia
Japan
Philippines
N. America
Equator
S. America
Australia
New Zealand
More than half of the world's active volanoes are found in the Ring of Fire. This is a region of volcanic and earthquake activity located around the rim of the Pacific Ocean.

A volcano is caused by pressure within the Earth forcing melted rock to the surface. The molten rock with dissolved gases lies in a ***magma chamber*** deep beneath the opening in the crust. When pressure builds up, the magma and gases escape through a tube-like passage called a ***vent***. Sometimes ***secondary vents*** develop farther down from the ***main vent***. When the volcano erupts, it spews gas, dust, ash, and rocks into the air. When the magma reaches outside Earth's crust, it is called ***lava***. The hot lava flows down the sides of the volcano. Sometimes, after an eruption, the top of a volcano collapses, forming a basin-like depression called a ***crater***.

Kinds of Volcanoes

Volcanoes are grouped according to their shape and the type of material of which they are composed. The shape of a volcano depends on the type of eruption that created it.

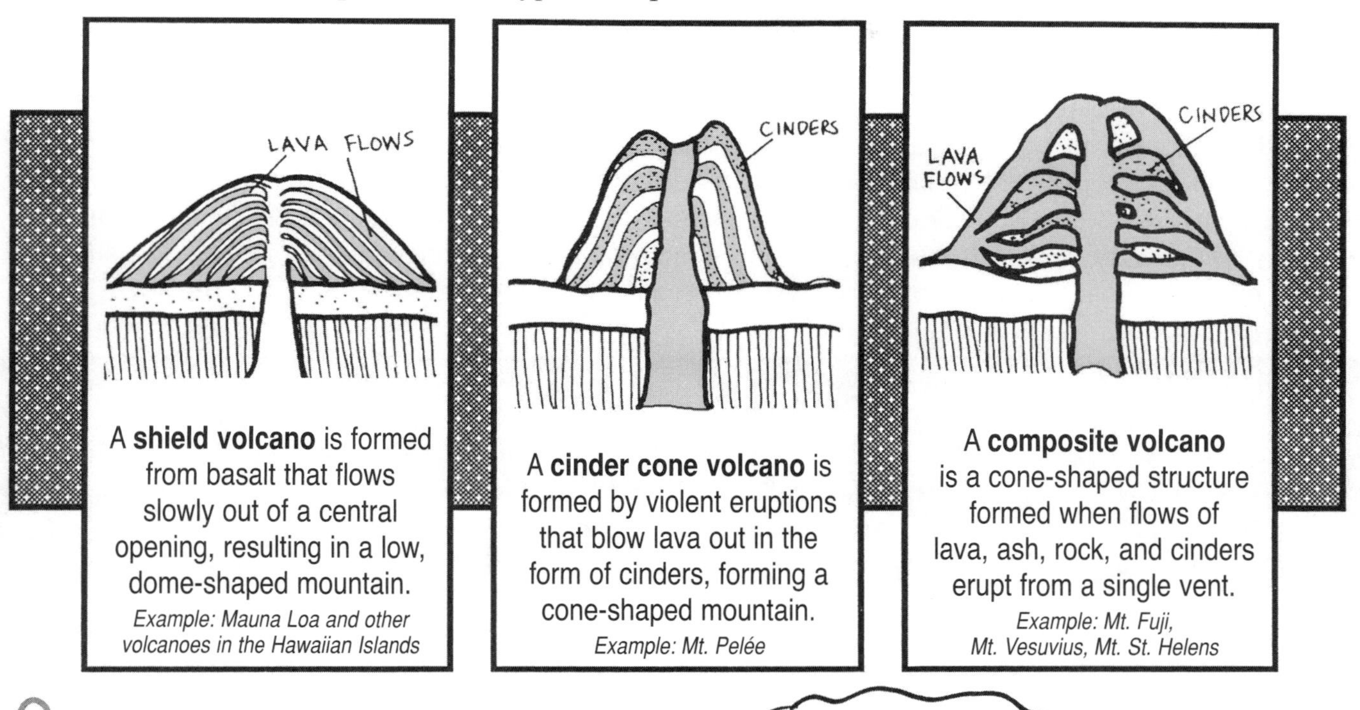

A **shield volcano** is formed from basalt that flows slowly out of a central opening, resulting in a low, dome-shaped mountain.
Example: Mauna Loa and other volcanoes in the Hawaiian Islands

A **cinder cone volcano** is formed by violent eruptions that blow lava out in the form of cinders, forming a cone-shaped mountain.
Example: Mt. Pelée

A **composite volcano** is a cone-shaped structure formed when flows of lava, ash, rock, and cinders erupt from a single vent.
Example: Mt. Fuji, Mt. Vesuvius, Mt. St. Helens

How to Make a Volcano

1. Find a piece of wood or cardboard to use as a base.
2. Use clay or papier mâché to create a mountain. Shape the mountain around a small glass jar.
3. When the model is dry or hardened, paint and shellac it.
4. Put $\frac{1}{2}$C baking soda into the jar in the center.
5. Gradually pour in vinegar, and watch the volcano erupt!

An **active** volcano erupts constantly.
An **extinct** volcano has been inactive for as long as any records have been kept.
An **intermittent** volcano erupts at regular intervals.
A **dormant** volcano is inactive, but could become active again.
Are you ready?
Fire away!
vinegar
baking soda

For more volcano information, visit the U.S. Geological Survey's volcano website: http://volcanoes.usgs.gov.

Minerals & Rocks

Rocks and minerals make up just about all of the Earth. Minerals are formed from combinations of elements. Rocks are made from combinations of minerals.

Minerals

A ***mineral*** is a naturally occurring, inorganic solid. Each mineral has a particular chemical make-up arranged in a crystalline pattern. There are over 2,000 minerals that have been identified. A few minerals consist of a single element. The others are combinations of elements. Almost all minerals are made from these elements: oxygen, silicon, aluminum, iron, magnesium, sodium, potassium, and calcium. A mineral generally contains the same kinds of atoms in the same proportions, arranged the same way. The structure of most minerals fits into one of these six crystal systems.

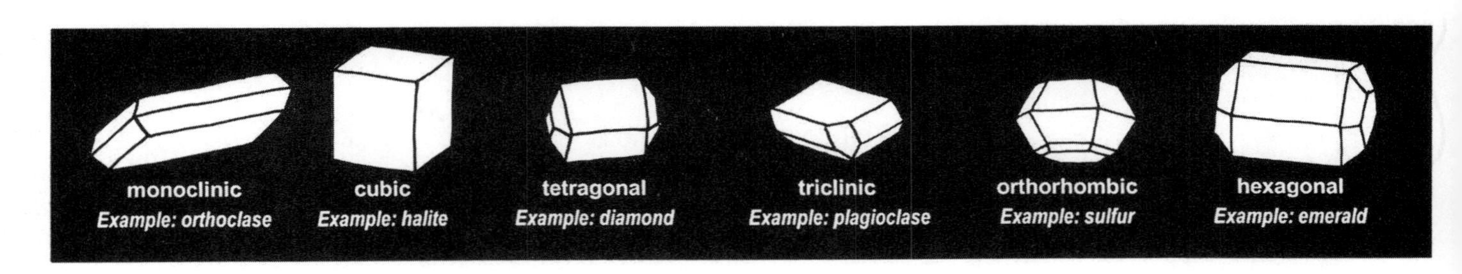

What did the little rock say to the big rock?
Stop taking me for granite!

Gems

A ***gemstone*** is a mineral that is especially rare and stands apart from other minerals because of its beauty. Color, luster, radiance, and hardness are properties that make a mineral valuable as a gem. Diamonds, rubies, opals, emeralds, and sapphires are all gemstones.

Identification of Minerals

Minerals are identified by their physical properties. You can learn to recognize many common minerals by paying attention to or testing these characteristics.

1. **Color**
2. **Form** is the shape or texture (crystal shapes, grainy fibers).
3. **Specific gravity** is the mass of a mineral in relation to water.
4. **Hardness** is the mineral's resistance to being scratched. A harder mineral scratches a softer one. (See Moh's Hardness Scale, page 107.)
5. **Luster** is the way light is reflected from the surface of the mineral. The luster can be *metallic* (shiny) or n*onmetallic* (dull, glassy or pearly).
6. **Streak** is the color a mineral makes when it is rubbed across of a piece of porcelain tile.
7. **Breakage** is the way a mineral breaks. *Cleavage* is breakage along smooth, flat planes. *Fracture* is irregular breakage (rough, jagged, or curved).
8. **Special characteristics** (tastes, odors, sounds, reactions with other substances)

Physical Properties of Some Common Minerals

The Cullinan Diamond is a South American diamond found in 1905. At 3106 carats, it is the largest diamond ever found.

$9.8 million was the price paid for the most expensive diamond sale on record. The rough 225-carat gem was sold in 1989.

Metallic Luster

Mineral	Color	Streak	Hardness	Crystals
graphite	black to gray	black to gray	1–2	hexagonal
silver	silvery, white	gray to silver	2.5	cubic
galena	gray	gray to black	2.5	cubic
gold	pale golden-yellow	yellow	2.5–3	cubic
copper	copper red	copper red	3	cubic
magnetite	black	black	6	cubic
pyrite	light brassy yellow	greenish black	6.5	cubic

Nonmetallic Luster

Mineral	Color	Streak	Hardness	Crystals
talc	white, greenish	white	1	monoclinic
gypsum	colorless, gray, white	white	2	monoclinic
sulfur	yellow	yellow to white	2	orthorhombic
halite	colorless, red, white, blue	colorless	2.5	cubic
calcite	colorless, white	colorless, white	3	hexagonal
fluorite	colorless and many colors	colorless	4	cubic
hornblende	green to black	gray to white	5–6	monoclinic
feldspar	gray, green, white	colorless	6	monoclinic
quartz	colorless, colors	colorless	7	hexagonal
garnet	yellow-red, green, black	colorless	7.5	cubic
topaz	white, pink, yellow, blue, colorless	colorless	8	orthorhombic
corundum	colorless, blue, brown	colorless	9	hexagonal
diamond	colorless, many colors	colorless	10	octagonal or hexagonal

The beautiful green emerald is one of the most rare and valuable of all gems. Emeralds are found in Brazil, Colombia, Siberia, Madagascar, and the U.S. Its composition is $Be_3Al_2(Si_6O_{18})$.

Moh's Hardness Scale

Mineral	Hardness	Hardness Test
talc	1	softest, can be scratched by fingernail
gypsum	2	soft, can be scratched by fingernail but not by talc
calcite	3	can be scratched by penny
fluorite	4	can be scratched by steel knife or nail file
apatite	5	can be scratched by steel knife or nail file, but not easily
feldspar	6	knife cannot scratch it; it can scratch glass
quartz	7	scratches glass and steel
topaz	8	can scratch quartz
corundum	9	can scratch topaz
diamond	10	can scratch all others

Rocks

All rocks are formed from single minerals or combinations of minerals. The rocks of Earth fall into three main groups. They are classified according to the way they are formed.

Igneous Rocks

Igneous rocks are formed from hot, liquid Earth materials (magma or lava).

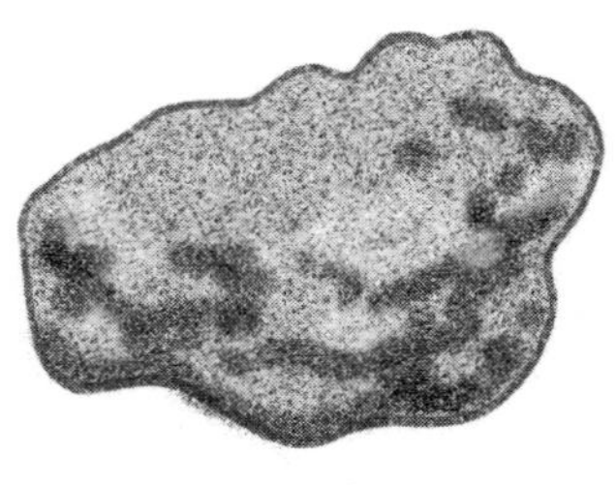

granite

Intrusive igneous rocks form when magma trapped below the surface cools slowly. The slow cooling leaves intrusive rocks with coarse textures.

Examples: olivine, peridotite, diorite, granite, orthoclase, quartz, basalt

Extrusive igneous rocks form when lava flowing on the surface cools quickly. The fast cooling results in rocks that are smooth and glassy or finely-grained.

Examples: felsite, andesite, rhyolite, obsidian, pumice

Sedimentary Rocks

Sedimentary rocks are made of sediments (loose material on Earth's surface). When weathered rocks break up and are deposited on the ground, they eventually settle and harden in layers. Sedimentary rocks are made of series of layers.

shale

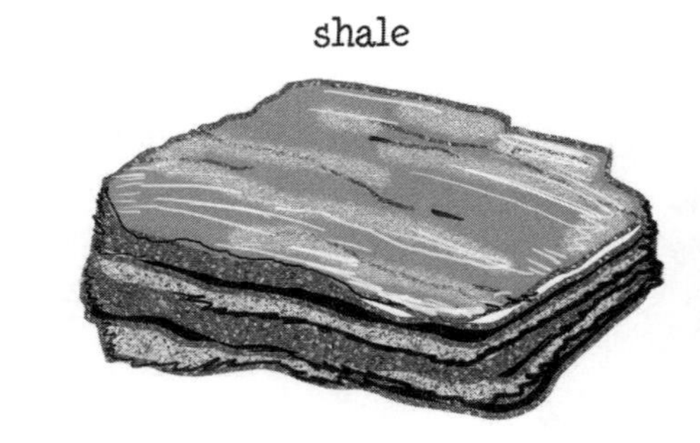

Clastics are sedimentary rocks made from fragments of rocks, minerals, and shells.

Examples: sandstone, siltstone, shale, conglomerate, breccia

coal

Nonclastics are sedimentary rocks formed from organic material or from chemical reactions.

Examples: calcite, limestone, peat, coal, flint, chalk, alabaster, rock gypsum

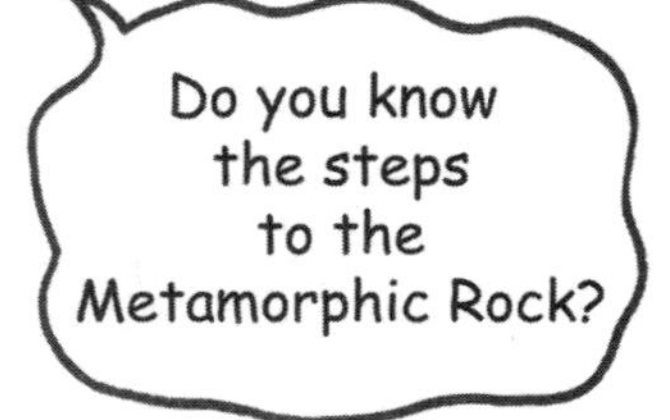

Do you know the steps to the Metamorphic Rock?
No, but I can sing the Rhyolite Rap while I do the Sandstone Slide.

Do you know the song, "Your Anthracite Heart"?

Is that the one recorded by Tommy Shale?

No, it's sung by the Lava-Ettes.

Metamorphic Rocks

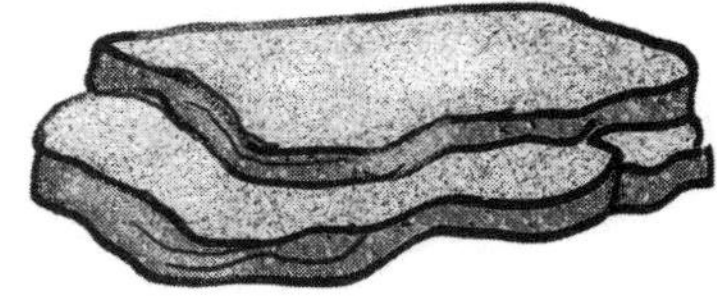
slate

Metamorphic rocks are rocks that have been changed by heat and pressure.

Foliated rocks have bands.
Examples: slate, phyllite, schist, gneiss

Nonfoliated rocks have no banding. They are usually massive.
Examples: quartzite, anthracite, graphite, marble

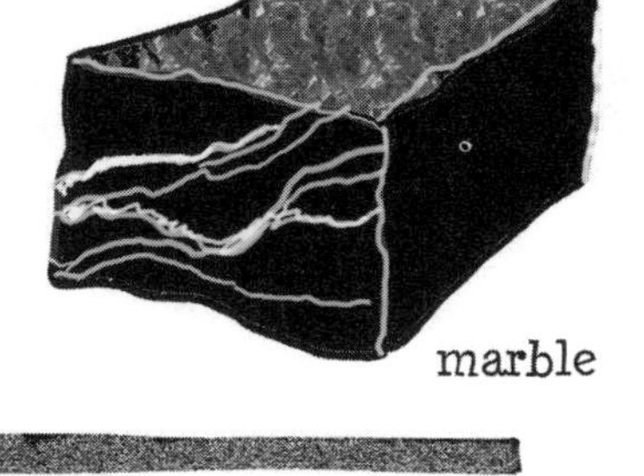
marble

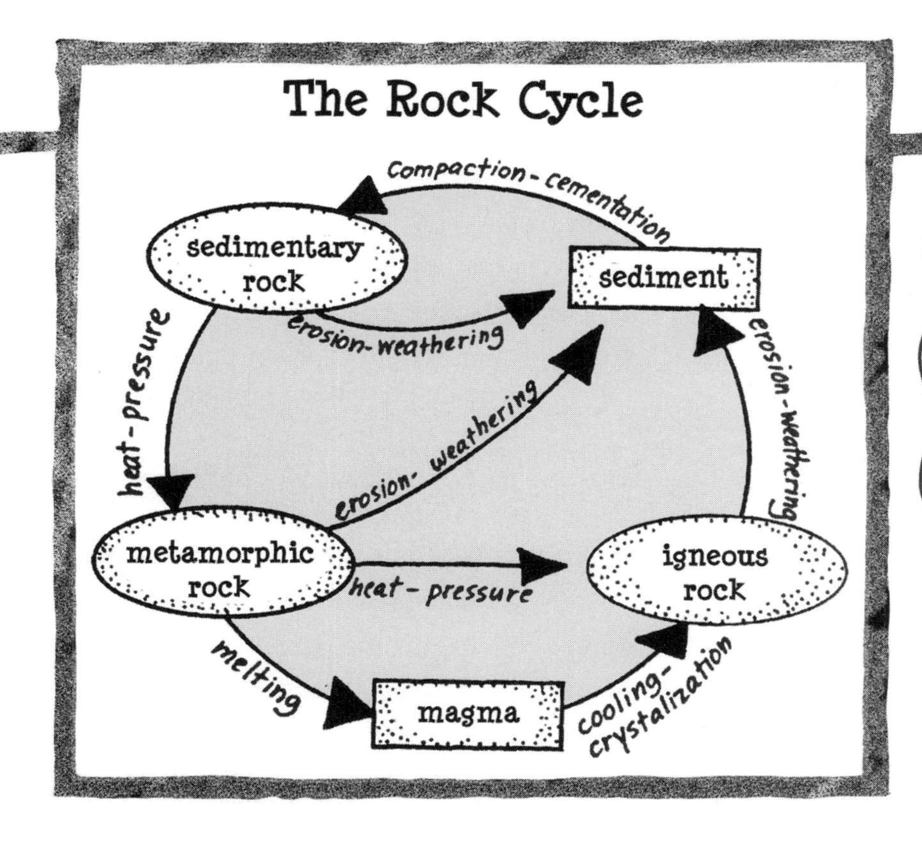

Rocks are constantly changing. Each kind of rock can be changed into the other two kinds. The rock cycle shows changes undergone by rocks.

Wonders of Earth's Surface

The surface of Earth is a wonderful collection of peaks and troughs, dry lands and wild seas, breathtaking views and curious formations. This brief glossary reviews some of Earth's major features. Match the numbers to see the features on the map (on the next page).

archipelago — large group or chain of islands

bay — part of a large body of water that extends into the land

butte — small, flat-topped hill; smaller than a mesa

cape — part of a coastline that projects into the water

canyon — deep, narrow valley with steep sides

channel — narrow strip of water between two land bodies

cliff — steep rock face

delta — land formed by deposits at the mouth of the river

desert — a dry area of land receiving little precipitation

fjord — deep narrow inlet of the sea between steep cliffs

foothills — hilly area at the base of a mountain range

glacier — large sheet of ice that moves slowly over a land surface

hill — rounded, raised landform, lower than a mountain

island — body of land completely surrounded by water

isthmus — narrow strip of land (bordered by water) joining to larger bodies of land

lake — body of water completely surrounded by land

mesa — high, flat landform rising steeply above surrounding land

mountain — high, rounded, or pointed landform with steep sides

mouth — place where a river empties into a body of water

peak — pointed top of a mountain or hill

peninsula — body of land nearly surrounded by water

plain — large area of level or gently rolling land

plateau — area of high, flat land; larger than a mesa

river — large stream of water that flows across land and empties into a larger water body

strait — narrow waterway or channel connecting two bodies of water

valley — V-shaped depression between mountains or hills

volcano — mountain created by volcanic action

waterfall — flow of water falling from a high place to a low place

FEATURE FEATURES

The plateau of Tibet, 850,000 square miles in area, is called the "Roof of the World." This is half the size of the U.S.A.

The Aleutian Islands form an archipelago that stretches 1,100 miles from the tip of Alaska.

Greenland is the world's largest island. It covers 840,000 square miles.

Arabia is the largest peninsula. It is a million square miles in area.

Surface Features
peak
volcano
mountain
fjord
glacier
plateau
foothills
canyon
mesa
butte
delta
river
archipelago
desert
mouth
strait
cliffs
hills
channel
valley
cape
hills
island
peninsula
isthmus
bay
plain
lake
waterfall

Surface Changes

Conditions on Earth's surface are constantly changing the rocks and minerals that cover most of it. Some of the changes are slow; some happen very fast. All of them alter the composition, form, or location of materials on Earth.

What's the Difference?

Weathering

is the wearing down or changing of rocks at or near the surface. Physical and chemical weathering eventually break down all rocks. Material resulting from weathering is called ***sediment***.

Erosion

is the process by which weathered particles are moved to another location. Agents of erosion pick up particles and leave them somewhere else. As rock particles are moved, they can become agents of erosion themselves. Rocks or sand scrape and scour the land's surface as they are carried along by wind, a moving stream, or ice.

The agents of erosion are gravity, wind, ice, and moving water. Sometimes people or their equipment are agents of erosion too!

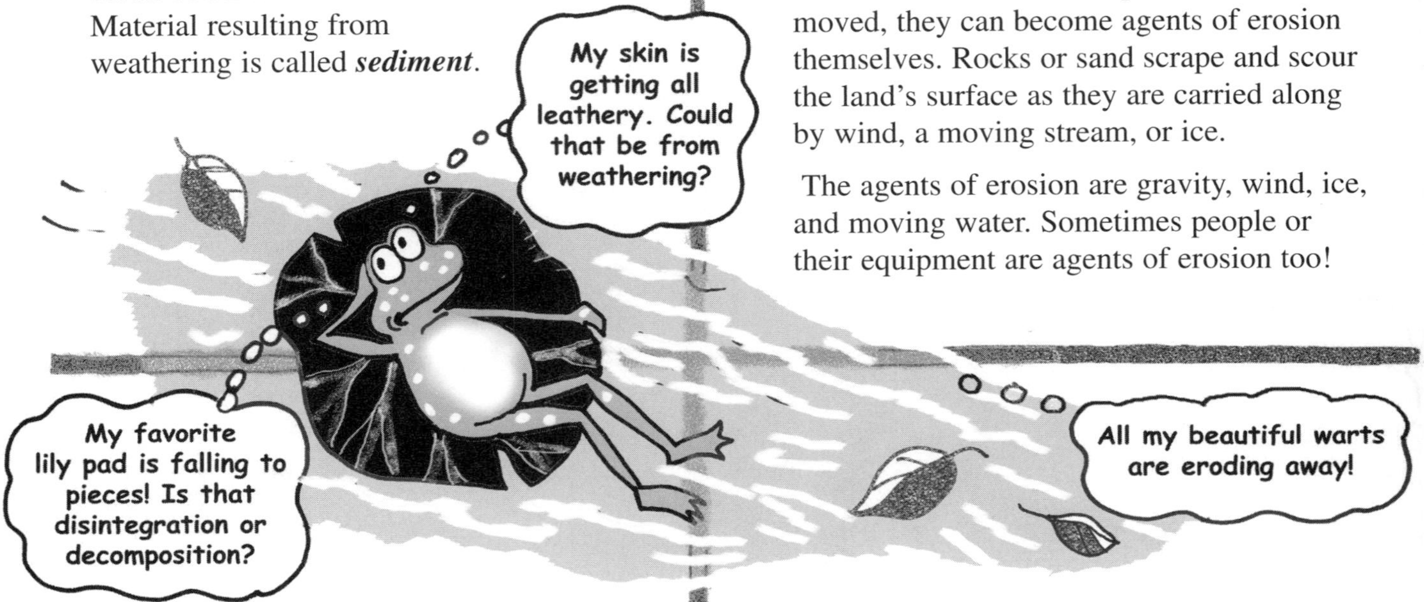

My skin is getting all leathery. Could that be from weathering?
My favorite lily pad is falling to pieces! Is that disintegration or decomposition?
All my beautiful warts are eroding away!

Disintegration

is physical weathering. This process breaks rocks down into smaller particles such as pebbles, dirt, sand, or other sediments. Wind, moving water, ice, gravity, plant roots, temperature changes, and the actions of people all contribute to disintegration. Disintegration happens faster in climates where there is freezing and thawing.

Decomposition

is chemical weathering. In this process, minerals in the rocks are dissolved by water or react with substances in the air. Decomposition changes the composition of the materials. Warm temperatures speed up the process of decomposition.

Gravity: an Agent of Erosion

I get in a real slump when a landslide falls downhill and fills my home with mud.

Gravity is the force pulling everything downward. This pull causes movement of particles or loose material down a slope. The material that piles at the bottom of a slope is called ***talus***.

Creep occurs when a mass of material moves very slowly downslope.

Slump occurs when layers of rock slip downslope leaving a curved scar.

A ***rockfall*** occurs when large masses of rock fall downslope quickly.

A ***landslide*** occurs when large amounts of material move quickly downslope.

A ***mudflow*** occurs when masses of debris and dirt mixed with rain slide downslope quickly.

Wind: an Agent of Erosion

It's hard being little in a wind storm!

Wind (moving air) is a powerful agent of change on the face of the Earth. Its force picks up loose materials and moves them around, piling them in some places and whipping them around against rocks like scouring brushes. Wind is a major force of erosion in dry areas like deserts and areas where sand is plentiful, like shorelines.

Deflation is the process by which wind removes loose material from surfaces. In deflation, wind may pick up dirt, dust, sand, or topsoil. This is the cause of ***dust storms*** and ***sandstorms***. Fine dust deposited by wind is called ***loess***. Sand carried by the wind will be dropped when it meets an obstacle such as a bush or clump of grass. The result is a ***sand dune***. When the wind sweeps all the material away in an area, the bare rock surface is called ***desert pavement***. Sometimes wind erodes soil and rock down to a level where water is present. Vegetation takes root and an ***oasis*** is formed.

Abrasion is the process in which particles in the wind scour and scrape against other surfaces. Rocks, cliffs, wood surfaces and other materials exposed to constant wind can become polished or pitted with this "sandblasting" action.

An **erg** is a sea of sand–a massive stretch of sandy desert and dunes. The largest one is the Grand Erg Oriental in northern Africa. (76,000 sq. miles)

Dunes

A ***sand dune*** (drift or pile of sand) forms when the sand-carrying wind meets an obstacle. Hitting an obstacle causes the wind to slow, and the sand is dropped. Sometimes vegetation begins to grow on dunes and slows the sand's movement. If vegetation does not settle into a sand dune, the dune will continue to move with the wind. Dunes can build up as high as 100–1,000 feet. They take different shapes and patterns, depending on the direction and intensity of the wind, the amount of vegetation, and the amount of sand.

Ice: an Agent of Erosion

Ice covers over 10% of Earth's surface permanently. Almost 75% of Earth's total supply of fresh water is frozen. The moving ice in glaciers and the ice that forms in cracks and crevices has great power to change Earth's surface.

Glaciers

A ***glacier*** is a large mass of ice in motion. A glacier forms when snow falls and doesn't melt. Over time, layers of snow accumulate into huge masses. The snow becomes compressed and heavy, and is packed into clear, blueish ***glacial ice***. The great weight causes slight melting at the bottom. The glacier slides and spreads along on this watery surface called ***meltwater***. A glacier moves very slowly, but as it does, its weight and the debris it carries become powerful causes of erosion. The glacier slides along cracking, souring, and gouging the rock surface below.

Glacial flow is the movement of glacial ice.

A ***valley glacier*** (or alpine glacier) forms at high elevations.

A ***piedmont glacier*** is a valley glacier that spreads out onto a plain.

A ***continental glacier*** is a great mass of ice found near one of Earth's poles.

Icebergs are large blocks of ice that break from a glacier to float in the sea.

The Antarctic's Lambert Glacier is the world's longest glacier. (250 miles long)

Glacial Erosion

Glaciers change the land as they move, scooping out formations such as cirques, horns, and hanging valleys, and leaving piles of debris.

Abrading is the scouring of rock surface that happens as a glacier moves over a surface carrying rough debris with it. Long scrapes, or striations, are left in the rock surface.

Plucking is the process of picking up fragments (large and small) and carrying them along.

The Antarctic's Lambert Glacier is the world's longest glacier. (250 miles long)

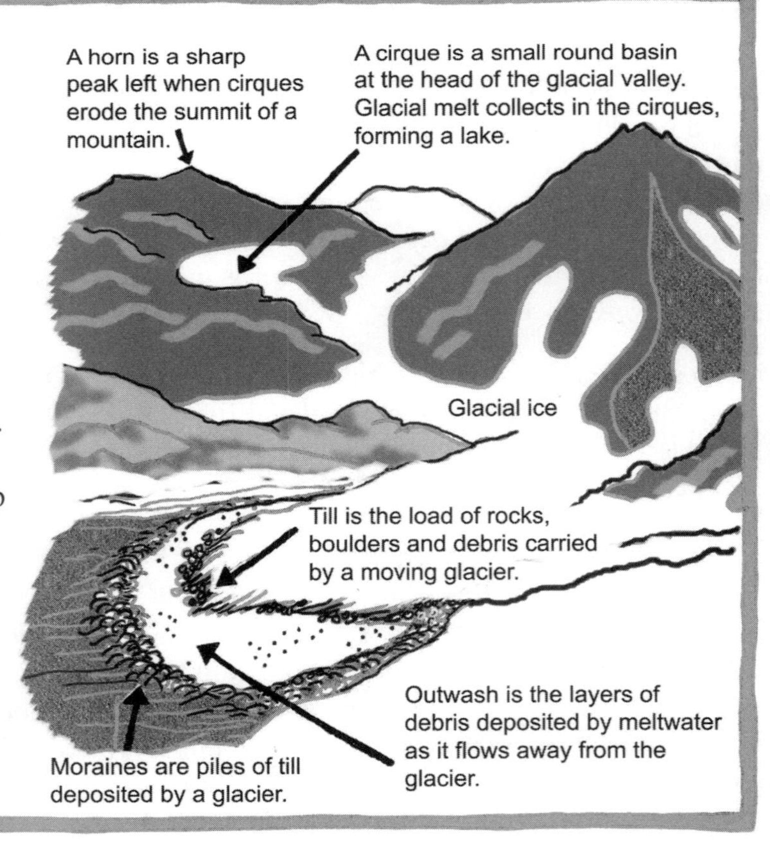

A horn is a sharp peak left when cirques erode the summit of a mountain.
A cirque is a small round basin at the head of the glacial valley. Glacial melt collects in the cirques, forming a lake.
Glacial ice
Till is the load of rocks, boulders and debris carried by a moving glacier.
Outwash is the layers of debris deposited by meltwater as it flows away from the glacier.
Moraines are piles of till deposited by a glacier.

Moving Water: an Agent of Erosion

Moving water is ***the main agent of erosion***. Some of the most drastic and spectacular changes to Earth's surface are caused by runoff, rivers and streams, waterfalls, underground water, and the actions of ocean waves.

Runoff

Runoff is water from precipitation that flows across Earth's surface and eventually returns to streams, rivers, and the ocean. As it flows, this water moves soil and wears away rock. Runoff moves quickly on sloped land. On flat land, runoff tends to sink into the ground. Vegetation slows runoff because the roots of plants hold the water in the soil and keep soil from being carried away by the moving water. Areas with little soil or vegetation, or with impermeable rock close to the surface are particularly vulnerable to runoff.

1) In the water cycle, the heat from the sun evaporates some of the water from oceans, lakes, and rivers. Also, living things transpire water vapor into the air.
2) Air rises. Water vapor turns to liquid and forms clouds.
3) Water falls from the clouds as rain or snow.
4) Water soaks into the ground or returns to the bodies of water.

Rivers and Streams

Moving water has the power to carve through rock or carry amazing amounts of sediment and debris. Streams and rivers wear away rocks, move along dirt, rocks, and debris, and change the face of the Earth. The moving water does not cause all the erosion alone. The load carried along by the water does much of the work, scraping the underwater surfaces. When a river joins an ocean or lake, it slows down and drops much of the load it is carrying. This sediment builds up at the river's mouth and changes the land. When a river floods, the sediment it is carrying spreads out from its banks, making even more changes in the land.

Groundwater

Some of the water that falls as precipitation slowly seeps down through soil into cracks in the rocks. This underground water, called ***groundwater***, moves slowly beneath the surface through a series of pores in the rocks. Groundwater dissolves some rock and causes changes in Earth's lithosphere. It moves materials around, creating interesting formations and leaving deposits.

Oceans

Ocean waves carry the force of moving water. Often, their force is increased by the power of wind. Waves can move sand and rocks, or toss them against the shoreline. The force of waves constantly shapes and changes the coastline. When ocean water combines with fierce winds in a storm, great damage can be caused to coastal areas.

Earth's Waters

Rivers & Streams

Most rivers begin as runoff that flows through ruts in Earth's surface. The water that does not soak into the ground flows in trickles that join other trickles, gradually forming larger streams of water. Eventually enough streams combine to form rivers. The streams or smaller rivers that join into a large one are called ***tributaries.*** Other rivers are formed from natural springs that come out of the ground through cracks in Earth's surface. Still others start from lakes or melting glaciers.

Get Sharp Tip #8
The beginning of a river is called its source.
The world's shortest river is 120 feet long. The D River in Oregon flows into the Pacific Ocean.

River Systems

As runoff grows into streams that merge into a river, the water flows in a pattern called a ***drainage system***. The area drained by a river is the ***drainage basin***. Each system is made up of a network of streams that form a pattern. Drainage patterns differ depending on the kind of rock over which the river flows.

Erosion by Rivers

Rivers weather and erode the land as they flow along. The force of the water wears rock away. Rocks, boulders, sand, and dirt can be pushed along by the water or picked up and carried away, scraping the river bottom as they move. The water also dissolves some materials in the riverbed. The amount of erosion caused by a river depends on the velocity of the river, the slope of the land, and the size of the load it carries.

The **riverbed** is the path through which the river flows.
The **river's load** is the material carried by a river.
The **suspended load** is the sediment and debris picked up and carried along in the water.
The **bed load** is the material that is rolled along the river bottom because it is too heavy to be carried by the water.

Waterfalls

When water flows over an area of rock that has a layer of soft rock over a layer of hard rock, the soft rock will eventually wear away. If this results in the water flowing down a vertical rock surface, a ***waterfall*** is created.

Angel Falls in Venezuela is the highest waterfall in the world. It drops 3,212 feet.

River Deposits

Rivers don't hang on to all the sediment they carry. About 75% of the load a river has picked up is dropped before the river reaches the ocean. This sediment is dropped along the way in the beds of the rivers and streams that make up the river's drainage system.

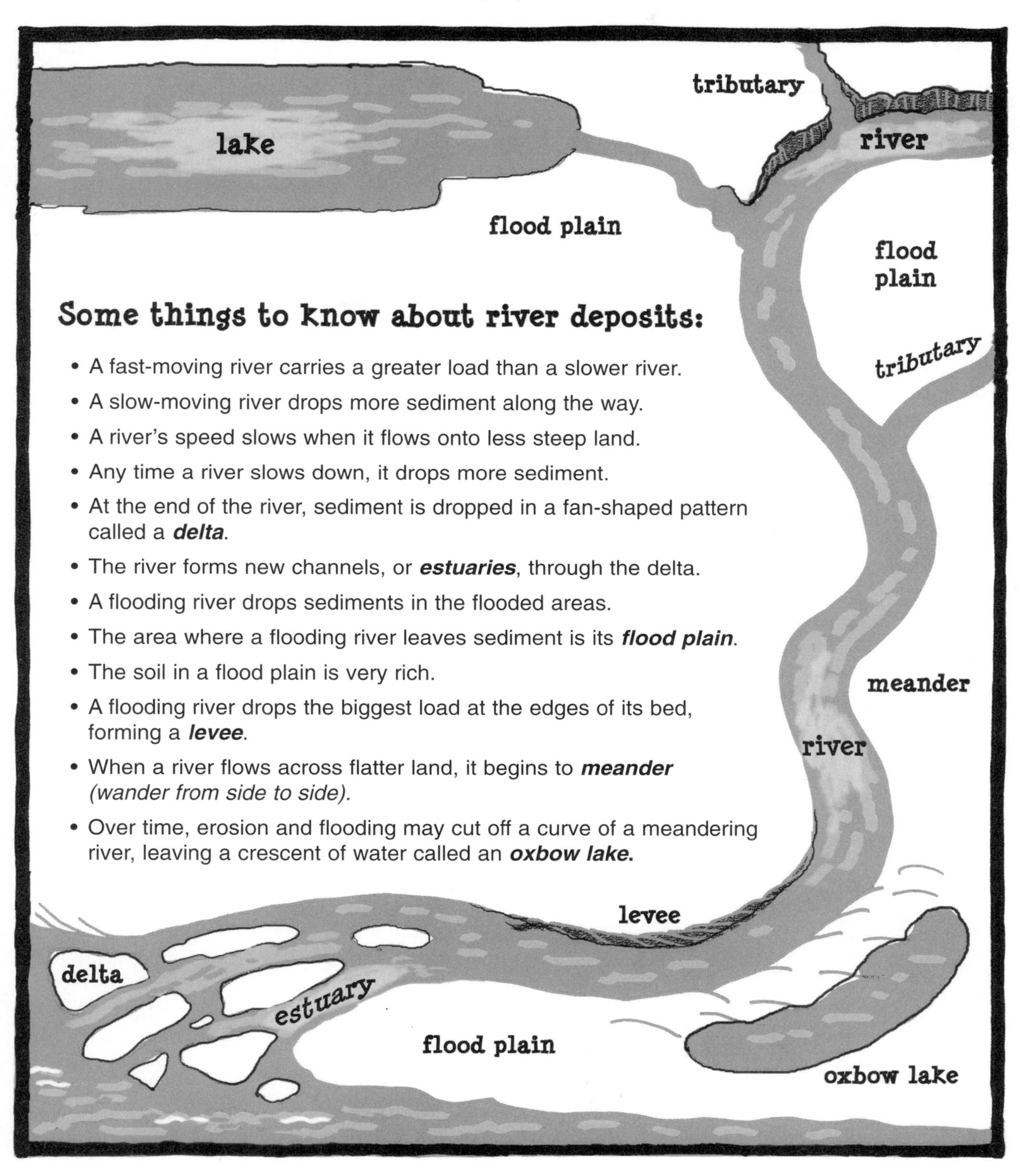

tributary
lake
river
flood plain
flood plain
Some things to know about river deposits:
• A fast-moving river carries a greater load than a slower river.
• A slow-moving river drops more sediment along the way.
• A river's speed slows when it flows onto less steep land.
• Any time a river slows down, it drops more sediment.
• At the end of the river, sediment is dropped in a fan-shaped pattern called a delta.
• The river forms new channels, or estuaries, through the delta.
• A flooding river drops sediments in the flooded areas.
• The area where a flooding river leaves sediment is its flood plain.
• The soil in a flood plain is very rich.
• A flooding river drops the biggest load at the edges of its bed, forming a levee.
• When a river flows across flatter land, it begins to meander (wander from side to side).
• Over time, erosion and flooding may cut off a curve of a meandering river, leaving a crescent of water called an oxbow lake.
tributary
meander
river
levee
delta
estuary
flood plain
oxbow lake

Groundwater

Some of the water on Earth's surface sinks into the ground, seeping into the soil and permeable rocks. This water becomes a part of a system of water beneath the ground, called ***groundwater.*** The action of groundwater causes many different deposits and natural formations.

Groundwater soaks down through soil and permeable rock until it reaches the layer of impermeable rock. The rocks just above the impermeable layers get soaked with water. The top edge of the of this area is called the ***water table***. Permeable layers of rock that are filled with water are called ***aquifers***. Gravity causes the movement of water through connecting pores in the aquifers.

Some groundwater returns to the surface in ***springs***. The water flows out of the ground through a natural opening in the ground. Wells can be drilled to remove water from the water table or aquifers. For most wells, the water must be pumped to the surface. ***Artesian wells*** need no pump. In these cases, the groundwater flows because of pressure from the heavy weight of a column of water in the aquifer.

Get Sharp Tip #9

Permeable rocks have tiny spaces between particles. In ***impermeable*** rocks, particles are so close together that there are no spaces.

There are over 2 million cubic miles of fresh water beneath Earth's surface.

Old Faithful in Yellowstone National Park is the world's best-known geyser. About every 76 minutes, it spouts gas and steam 120–150 feet into the air.
PARK RANGER

Hot Springs, Geysers, & Fumaroles

Hot Springs are pools or streams of water that are heated naturally by heat from beneath Earth's surface. Water seeps down through rocks that are hot because they are near to pockets of hot magma. The hot water rises back to the surface and gathers in pools or running streams of hot water.

Fumaroles are vents in the ground through which volcanic gases and steam escape. Fumaroles usually appear in regions of former volcanic activity. *Solfataras* are fumaroles that give off sulfurous gases. *Mudpots* are areas where fumaroles bubble up through mud.

Geysers form in places where water seeps into the ground to an area where the surrounding rock is hot. The water gets heated and forms steam. The steam pushes the water up through cracks in the Earth with such force that it explodes in a shower of steam and water. The water settles back into the Earth until the steam builds up again.

To learn more about groundwater, visit the kids' corner at www.groundwater.org

Erosion and Changes

Groundwater contains elements that can erode some of the softer rocks in and below the surface. The acid in the water dissolves away soft rocks like limestone, leaving the harder rock intact. This action creates some spectacular features.

Caves — When limestone is dissolved, some large openings are left in the harder rock. These are called caves or caverns. The process of cave-making takes thousands of years.

Natural Bridges — When a portion of a cavern roof collapses, but some of the roof remains, a natural bridge is formed.

Sinkholes — Water drains through cracks in limestone, gradually dissolving it. This causes funnel-shaped depressions called sinkholes. The ground above the sinkhole can give way since there is no support left underneath it. This can cause serious damage to roads or buildings above the sinkhole.

Stalactites & Stalagmites — Water moving underground can contain large amounts of dissolved substances. When groundwater drips from the roof of a cave, it evaporates and leaves calcium carbonate deposits that look like giant icicles. Structures called ***stalactites*** develop and hang from cave ceilings. The structures that build up from the floor of the cave are ***stalagmites***. Eventually, many stalactites and stalagmites join together into one column.

How to Make Your Own Stalactites and Stalagmites

You will need:

- 2 tall jars
- a spoon
- cotton string
- a flat plate
- a box of Epsom salts
- 2 small rocks
- hot water

1. Set the jars about a foot apart from each other.
2. Cut a piece of string long enough to reach between the jars, and into the bottom of each.
3. Tie a small rock around each end of the string.
4. Pour hot water to fill each jar about half full.
5. Mix salt or baking soda into both jars. Add it until no more will dissolve.
6. Soak the string in one jar of water. String it from one jar to the other, with the ends in the jars.
7. Set the plate underneath the center of the string.
8. Make sure no one bumps or disturbs the materials.
9. Watch carefully to see what happens as the water evaporates.

I'm going to demonstrate how to make stalactites and stalagmites. Follow these directions carefully.
My hobby is spelunking, so I've seen lots of these in real caves.
Did you know that the longest cave system in the world is in Kentucky? The Mammoth Cave system is 352 miles long.

Lakes, Ponds, & Swamps

> **LOCH TALK**
>
> Loch Ness is a lake in Scotland that is famous for something other than its size, depth, or water. It is well kinown because some people think a 50-foot long sea monster lives in the lake. This sea serpent is called the **Loch Ness Monster.**

Lakes

A ***lake*** is an area of water completely surrounded by land. Most lakes contain fresh water, though some are filled with salt water. Lakes are usually formed from the flow of a river. In most cases, a lake will eventually disappear. Rivers flowing out of the lake will drain it, or the water will evaporate, or the lake area will fill with vegetation and sediment.

Artificial Lakes

An artificial lake is one created by humans. Most artificial lakes are created when a dam is built across a river. The river water builds up behind the dam and floods the river valley. In most cases, a river is dammed for the purpose of producing electricity. The power of the flowing river is used to run generators and make electricity at whatever rate is needed.

Have you ever wondered why is it harder to float in a freshwater lake than in salt water?
Here is the answer. Water with a high salt content has high density. The water molecules are closer together. This gives the water a stronger base for holding me afloat!
I'd be having an easier time floating in the Great Salt Lake.

Ponds

A ***pond*** is a small, shallow body of water. In most ponds, the sunlight can reach the bottom of the pond. This allows plant life to spread across the entire bottom surface.

> **The largest swamp in the world is in Brazil. The Grand Pantanal Swamp covers 42,000 square miles.**

Swamps

A ***swamp*** is a shallow body of water, usually poorly drained, found in a low-lying area. The water is filled with plant material that has fallen into the lake and decayed into thick, mucky layers of peat.

> **The Everglades is the largest swamp in the United States. It covers 5,000 square miles.**

Oceans

Over three-fourths of the Earth's surface is covered with water (liquid or frozen). About 98% of this water is found in the world's oceans.

An *ocean* is a large body of salt water. Sometimes oceans are called *seas*, although a sea is generally thought of as a body of salt water smaller than an ocean.

There are five oceans: the Arctic, Atlantic, Pacific, Indian, and Southern (or Antarctic). The oceans include smaller bodies of water around their edges, such as gulfs and bays.

The combined volume of all five oceans is 317 million cubic miles.

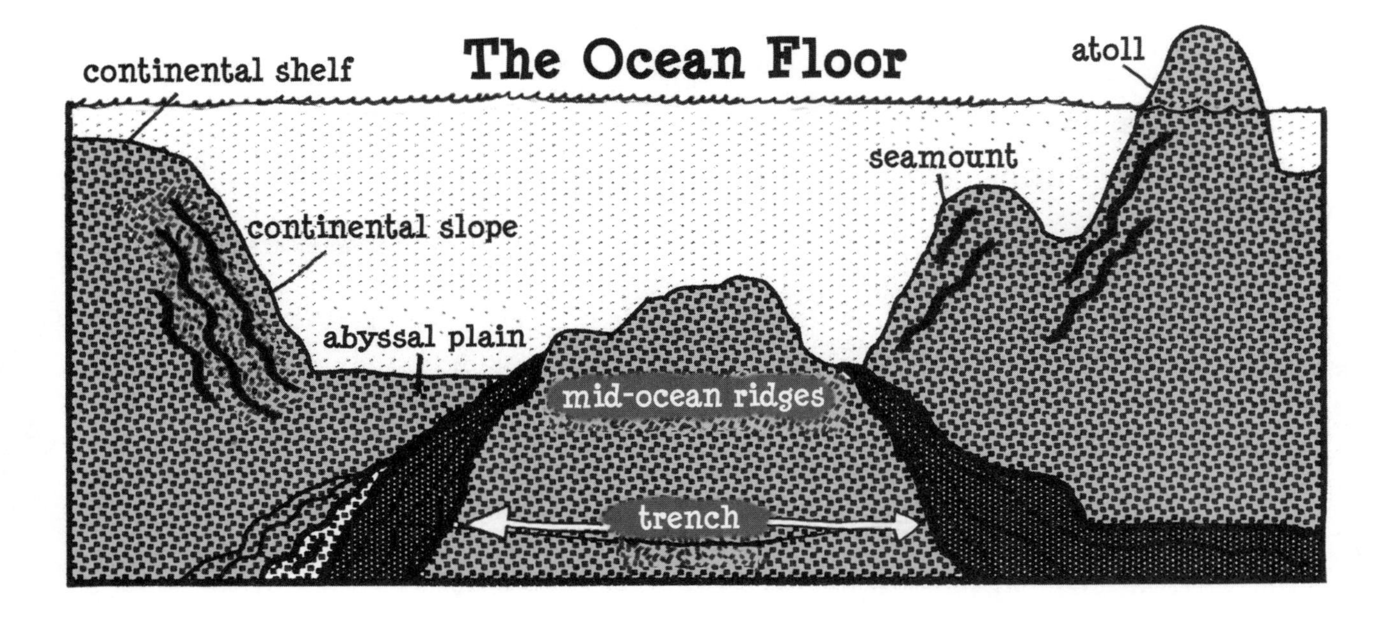

The bottom of the ocean is as varied as other parts of Earth's crust. It contains mountains, valleys, plains, crevasses, ridges, and volcanoes.

continental shelf — underwater land at the edges of the continents, from the shoreline to about 600 feet deep

continental slope — a steeper slope running downwards from the shelf to the ocean floor

abyssal plain — wide, flat area that makes up most of the ocean floor

mid-ocean ridges — mountain ranges on the ocean floor

trench — a long, narrow crevasse in the ocean floor

seamount — a mountain with a peak below the water surface

atoll — a mountain on the ocean floor that breaks through the water surface

The deepest spot in an ocean is the Mariana Trench in the Pacific Ocean, near Guam. It is 35,840 feet deep (about 6 miles).

Water Movements

The water in the oceans is always moving, even if it looks still at the surface. Winds create waves or currents. Earthquakes create giant waves. The gravitational pull of the Sun and Moon produce falling and rising movement called ***tides***. Beneath the surface, circulation (vertical movement) of water is caused by wind patterns, Earth's rotation, and differences in temperature and salt content.

Currents

Surface currents are caused by winds blowing across the surface of the water. The wind whips water into motion, forming horizontal currents that flow in circular patterns. Wind currents move the upper levels of the ocean water, sometimes as far down as 600 feet or more. Earth's rotation adds to the circular pattern of the currents. The currents are interrupted or affected by the continents.

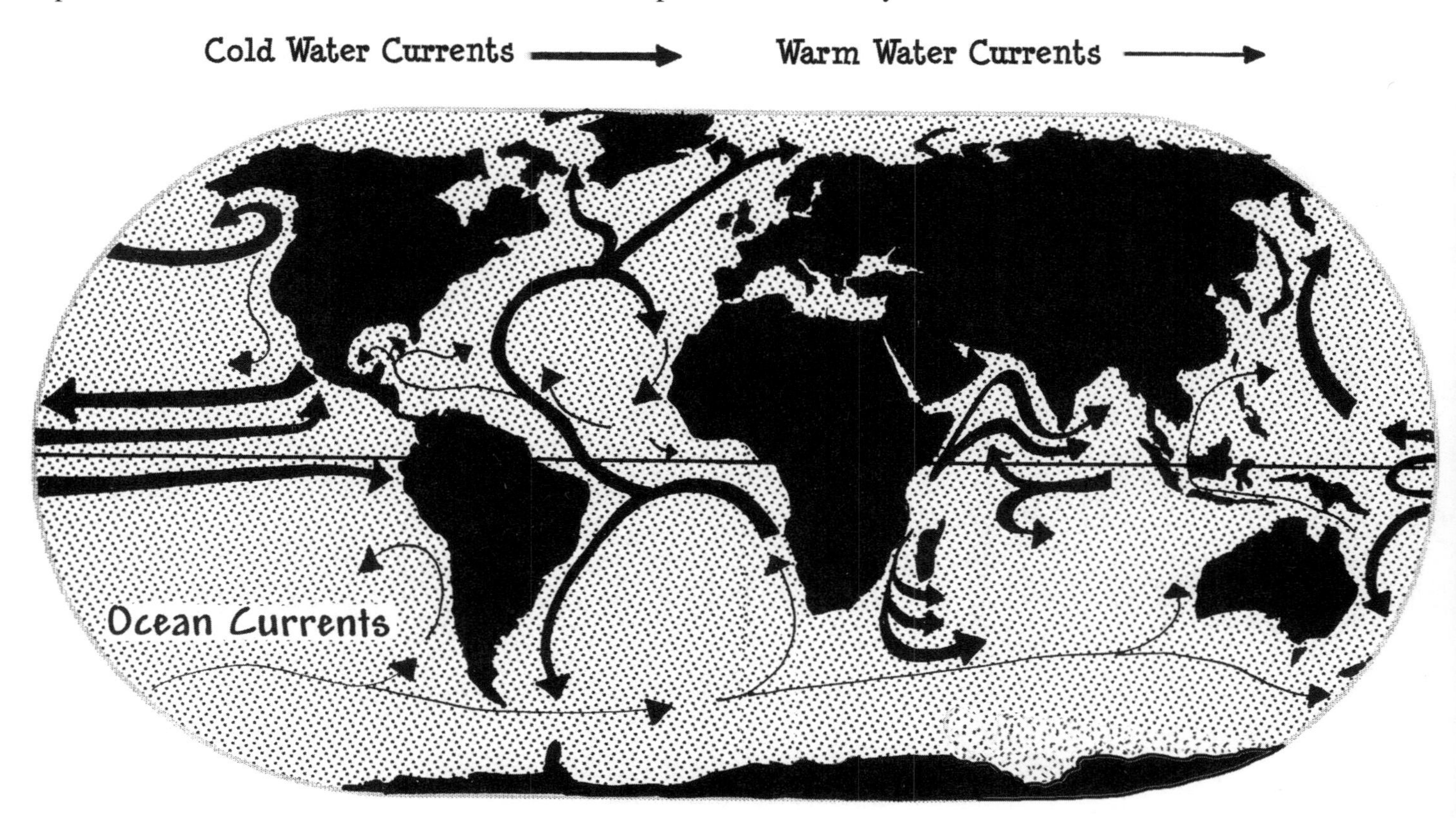

Ocean Currents

Deep Water Currents (also known as ***density currents*** or ***thermohaline currents***) result from differences in the temperature and chemical content of the water. ***Salinity*** is the measure of dissolved solids or chemicals in the water, much of which is salt. Water with high salinity is denser than water with lower salinity. Cold water is also denser than warm water. Therefore, cold water is heavier and usually sinks, causing warmer water to rise upwards. Deep water currents are caused by movement of high density water into areas of lower density water.

Get Sharp Tip #10

An ***upwelling*** is an unusual case where a pocket of cold water, rich in nutrients, rises up to the surface. Upwellings create good fishing areas, because the nutrients attract fish.

Waves

Waves are ocean movements in which water rises and falls. Wind, earthquakes, or tides cause waves. Within a wave, water particles are moving in a circle. The height of a wave is equal to the diameter of the circle made by the particles moving in the wave.

A wave can be described by these features:

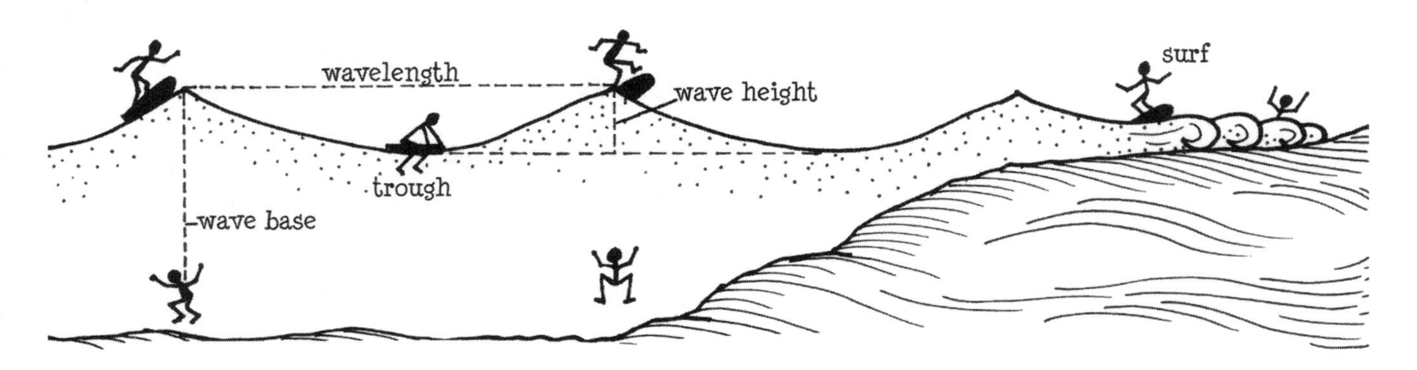

Crest – the highest point of a wave

Trough – the lowest point of a wave

Wave height – the vertical distance from a wave's crest to the trough

Wavelength – the horizontal distance between two wave crests

Wave base – the depth of water equal to one- half the wavelength

Wave Period – the time it takes two consecutive crests to pass a point

Surf – breaking waves along the shore

Shallow water wave – a wave in water shallower than one-half its wavelength

Deep water wave – a wave in water deeper than one-half its wavelength

Tsunami – a wave caused by an earthquake

Tides

Tides are shallow water waves caused by the interaction among the gravity of Earth, Sun, and Moon. The Moon's gravitational force causes ocean water to bulge toward it, resulting in a ***high tide***. Rotational forces cause another bulge on the opposite side of the Moon. The depressions between high tides are ***low tides***. The ***tidal range*** is the difference between a high tide and a low tide. The relative positions of the Earth, Moon, and Sun affect the tides. During ***spring tides***, high tides are at their highest and low tides at the lowest. ***Neap tides*** are minimum tides. The tidal range is greatest during spring tides, and least during neap tides.

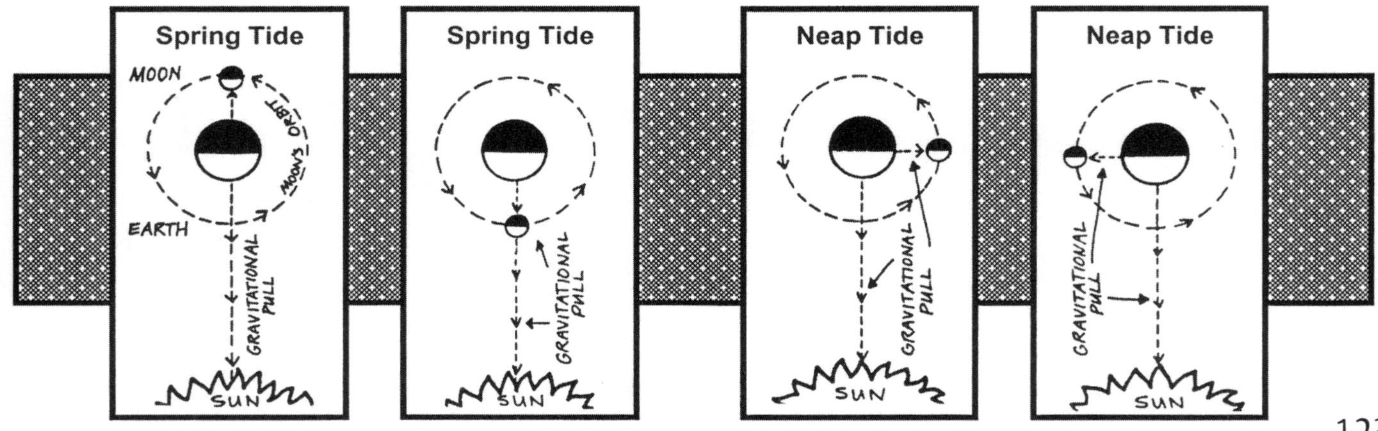

Shorelines

A ***shoreline*** is the boundary where the land meets the water of the ocean. The ***shore zone*** is the area along the shore that lies between the point of high tide and low tide. There is constant movement in shore zones. Actions of waves move sand and other sediments, causing many changes along the ocean shore.

Longshore currents (currents created when waves hit the beach at an angle) and ***rip currents*** (narrow currents that flow at a right angle to the shore) erode and change the shoreline.

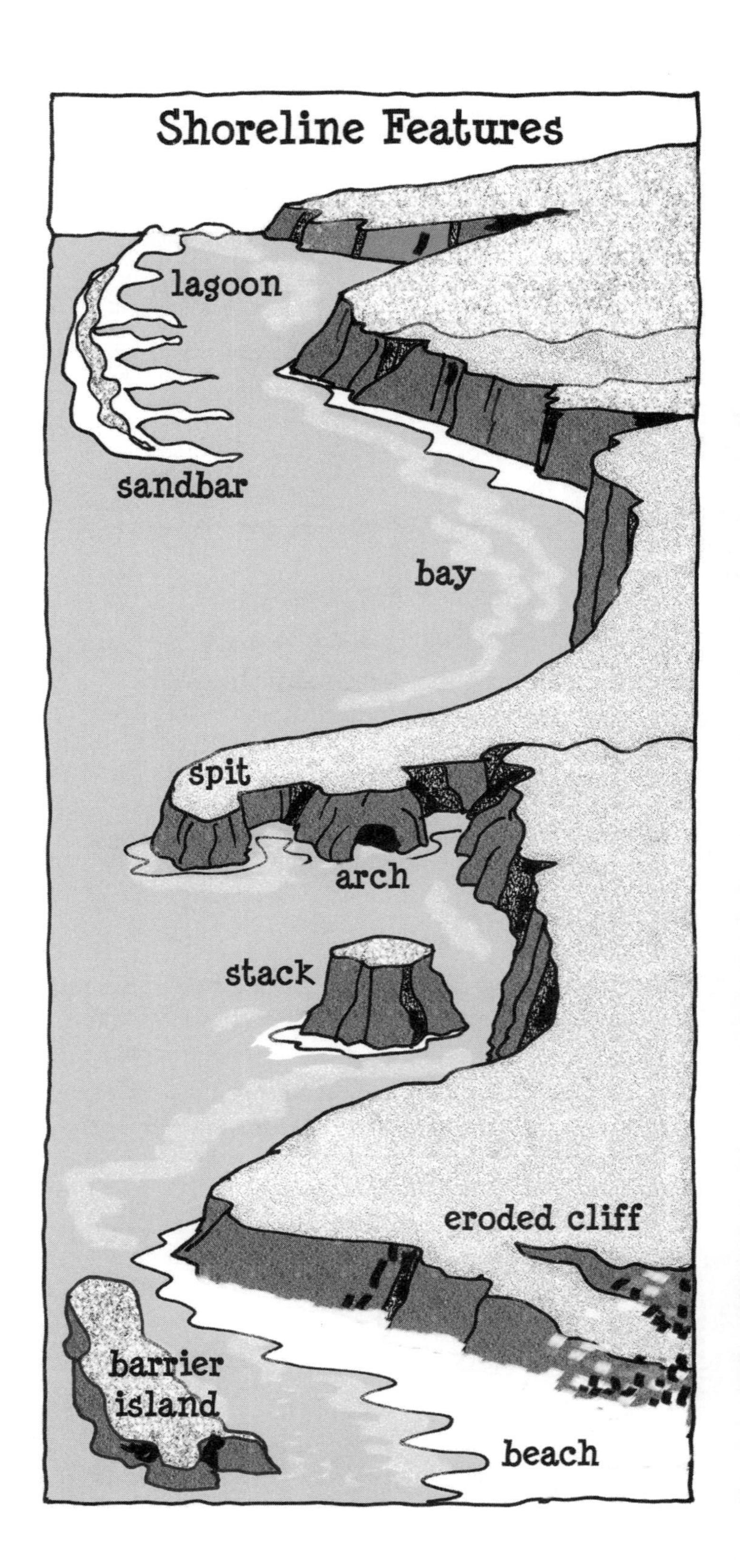

Shoreline Features

Beaches are deposits of sand and other fragments left along the shoreline boundary. Beaches extend to 100 feet or more above and below the shoreline.

A **sandbar** builds up where long-shore currents deposit sand and debris in deeper water parallel to the shore. The sandbar can be above water or covered by shallow water.

A **spit** forms along a curved shoreline where longshore currents drop loads of sand and debris extending across a bay.

A **bay** forms where part of the coastline is eroded.

A **lagoon** is a body of water cut off from the sea by a sandbar or reef.

Barrier islands develop from loads of sand and debris deposited parallel to the shore.

Arches and **stacks** are formations of resistant rock left standing after softer rock is worn away.

Earth's Atmosphere

The *atmosphere* is a blanket of air that surrounds the Earth. This envelope of air extends up to about 300 miles (480 kilometers) above Earth's surface. There is not a clear boundary where Earth's atmosphere ends. There are four layers in the atmosphere.

The **troposphere** (0–20 km) is the layer closest to Earth's surface. It contains most of the dust and gases (including water vapor) of the atmosphere. Clouds and weather occur in this layer. Jet stream winds occur at the very top boundary.

The **tropopause** (8–20 km) is the upper ceiling of the troposphere. It functions as a weather ceiling.

The **stratosphere** (20–50 km) is very cold, with temperatures of 0° to –50° C. The ***ozone layer*** is found in the stratosphere. Ozone is an important gas that absorbs most of the harmful ultraviolet radiation from the Sun.

The **mesosphere** (51-85 km) is the coldest layer of the atmosphere. Temperatures are as low as –100° C.

The **thermosphere** (above 85 km) is the uppermost layer. The air is thin, so energy from the Sun warms the air. The lower part, the ***ionosphere***, has electrically charged particles useful for transmitting radio waves. The ***exosphere*** extends from about 500 km into outer space. The air is so thin here, that some of its molecules escape into space.

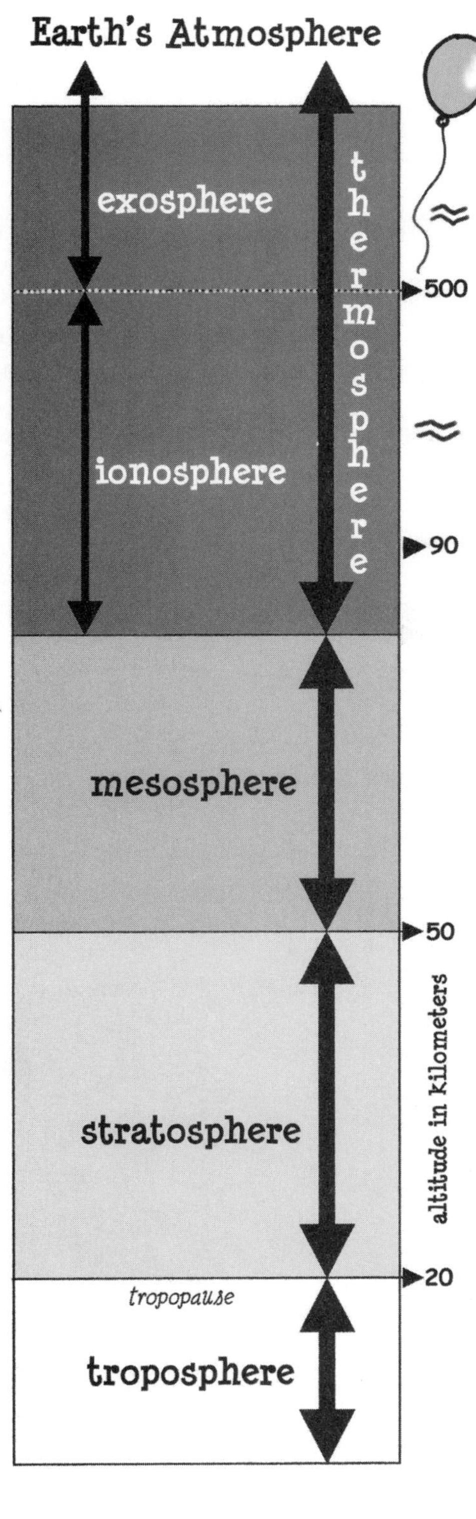

What IS air, anyway?
It is a mixture of gases. The air is 78% nitrogen, 21% oxygen, and 1% other gases.

Atmospheric pressure is the force of air pressing down on Earth's surface. At sea level, the atmospheric pressure is about 14.7 pounds per square inch. Air pressure varies in different places. A machine called a *barometer* is used to measure air pressure. Air pressure is affected by temperature and altitude.

Winds

Wind is moving air. Air always moves from areas of high atmospheric pressure to areas of low pressure. Air moves in great spirals from high to low pressure, creating giant wind systems around the Earth. Air also moves horizontally across Earth's surfaces. The differences in surface temperatures can also affect air movement.

Winds and Wind Systems

polar easterlies – These dry, cold air currents move from northeast to southwest over the polar zones (60° to 90° N latitude) of the Northern Hemisphere and from southeast to northwest over the polar zones of the Southern Hemisphere (60° to 70° S latitude).

westerlies – In the middle latitudes (30° to 60° north or south of the equator), the prevailing winds blow from west to east.

trade winds – These winds blow toward the equator from a latitude of about 30° north and south of the equator. They blow from northeast to southwest in the Northern Hemisphere and from southeast to northwest in the Southern Hemisphere.

Get Sharp Tip #11

The rotation of Earth causes air to move fastest at the equator. This creation of eastward winds just above and below the equator is called the **Coriolis Effect.**

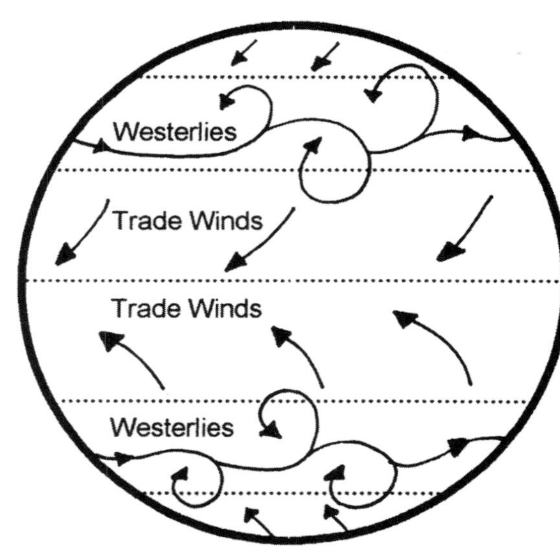

doldrums – At the equator, there is little horizontal movement of air. In this area, the air moves upwards, so it seems to be windless.

sea breeze – During the day, land absorbs heat from the Sun. The warm air rises and cool sea air blows onto the land.

land breeze – At night, the land loses its heat, and the warmer air over the water rises. Cooler land air moves out to sea to fill in the space.

jet streams – These narrow belts of fast-moving air flow in a westerly direction at the higher levels of the troposphere.

An instrument called an ***anemometer*** measures the speed, or velocity, of the wind. Velocity is measured in miles or kilometers per hour. A scale for measuring wind force was developed in 1806 by Sir Francis Beaufort. This is the scale, with an added description of the action caused by each level of wind.

Some Local Winds

Monsoons–stormy winds that bring heavy rain to southern Asia and northern Australia *(The rainy season, which lasts from about June to September, is called the Monsoon season.)*

Chinooks–warm, dry winds that flow down the side of a mountain range. *(These are called "foehns" in Europe.)*

Siroccos–winds that blow north from the Sahara Desert *(The winds pick up moisture over the Mediterranean Sea and bring warm rain to Europe.)*

Harmattans–cool, dry winds that blow west from the Sahara Desert *(They bring relief from the intense heat to the tropics of Northern Africa.)*

Typhoons–the name given to hurricanes in the northwest Pacific Ocean

Cyclones–the name given to hurricanes in the Pacific Ocean north of Australia and in the Indian Ocean

The Beaufort Scale of Wind Strength

Beaufort Number	Wind Name	Speed mph	Description
1	**light air**	1–3	wind direction shown by smoke drift; weather vanes inactive
2	**light breeze**	4–7	wind felt on face; leaves move slightly; weather vanes active; smoke does not rise vertically
3	**gentle breeze**	8–12	leaves and small twigs move constantly; flags blow
4	**moderate breeze**	13–18	dust and paper blow about; twigs and thin branches move
5	**fresh breeze**	19–24	small trees sway; whitecaps form on lakes
6	**strong breeze**	25–31	large branches move; telegraph wires whistle; umbrellas are hard to control
7	**moderate gale**	32–38	large trees sway; it is somewhat difficult to walk
8	**fresh gale**	39–46	twigs break off of trees; walking against the wind is very difficult; possible damage to property
9	**strong gale**	47–54	slight damage to buildings; shingles are blown off roof
10	**whole gale**	55–63	trees are uprooted; much damage to buildings
11	**storm**	64–75	widespread damage
12	**hurricane**	over 75	extreme destruction

Hurricanes are given men's or women's names. There are 6 sets of names for hurricanes. The lists rotate every 6 years.

Some names on the 2005 list are Arlene, Cindy, Franklin, Harvey, and Nate. Some 2006 names are Alberto, Leslie, Ernesto, Oscar, and Sandy.

Visit the FEMA website for more names: www.fema.gov.

Weather

Get Sharp Tip #12
When air masses meet, expect stormy weather at the front!

Weather is the state of the atmosphere at any one time. The factors involved in weather are temperature, air pressure, wind, and moisture.

Clouds & Precipitation

Clouds are groups of billions of tiny droplets of water in the air. They are formed when warm air rises and cools, or when warm air meets cold air. Since cooler air holds less water vapor than warm air, some of the vapor condenses. When the cloud becomes too full of water drops, water will fall from the cloud in different forms of ***precipitation*** (such as rain or snow).

Air Masses & Weather Fronts

When air slows over an area of land or water, a large body of air called an ***air mass*** may form. Air masses differ in temperature, atmospheric pressure, moisture content, and pattern of air circulation. In low-pressure air masses (called ***cyclones***), air moves counterclockwise toward the center. In masses of high pressure (called ***anticyclones***), air circulates clockwise out from the center.

When air masses meet, they form a boundary that keeps them separate. This boundary is called a ***front***. Different kinds of fronts can form, depending on the nature of the air masses that are meeting.

Some "Air-y" Facts

- Cold air sinks because it is denser than warm air.
- Dense air creates areas of high pressure.
- Warm air is less dense than cold air.
- Areas of low pressure are created when air warms.
- Warm air can hold more moisture than cold air.
- When air cools, it has to drop some of its moisture.
- Warm air rises, losing moisture as it cools.
- Cold air sinks and pushes warm air up.
- When the relative humidity of an amount of air reaches 100%, water vapor condenses into water.

Humidity is the ability of air to hold moisture. Air with high humidity has a high moisture content.

Dew point is the temperature at which water vapor condenses in a layer of air.

Relative humidity is a measurement of the amount of water vapor in the air in relationship to the total volume of the air.

Each different front that develops creates a particular kind of weather pattern.

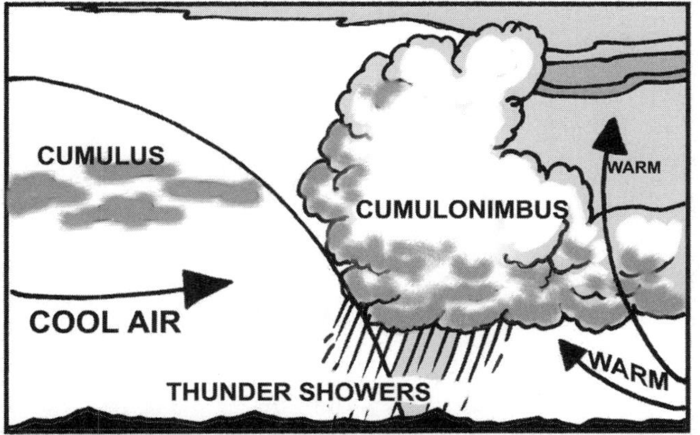

A **cold front** develops when a cold air mass invades a warm air mass. The heavier cold air sinks and slides under the warm air, pushing it steeply upwards. This causes cumulus and cumulonimbus clouds to form. Rainstorms or thunderstorms develop.

I'm waiting for a warm front to come by.

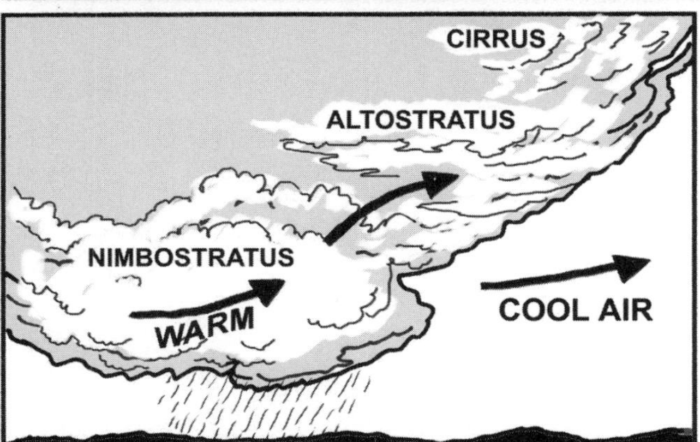

A **warm front** develops when a warm air mass invades a cold air mass. The less dense warm air slides up and over the cold air. Cirrus clouds, altostratus clouds, and nimbostratus clouds develop. Rain or snow often accompanies warm fronts.

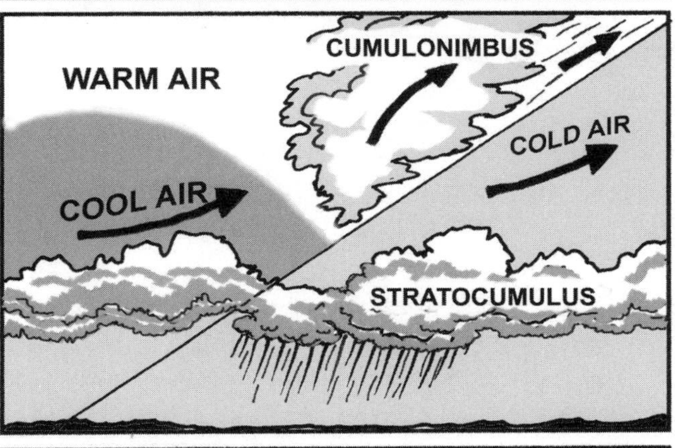

An **occluded front** develops when two masses of cold air meet. The cold air forces warmer air caught between the two fronts upwards. Cumulonimbus and stratocumulus clouds usually form. High winds and heavy rain or snow may result.

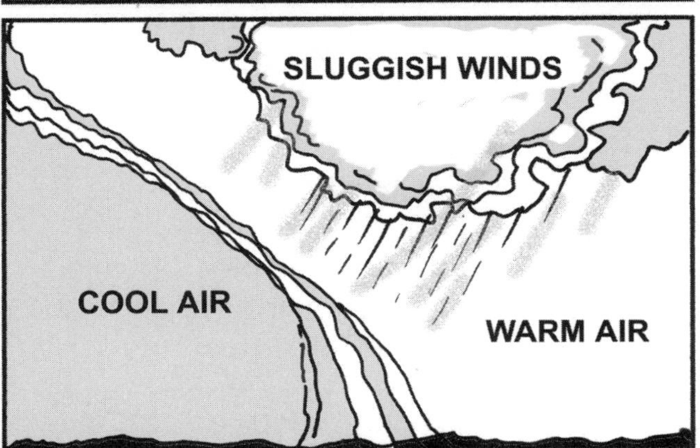

A **stationary front** develops when a cold front or a warm front stays in place for several days without invading another front. Clouds and precipitation often form at the boundary.

Types of Clouds

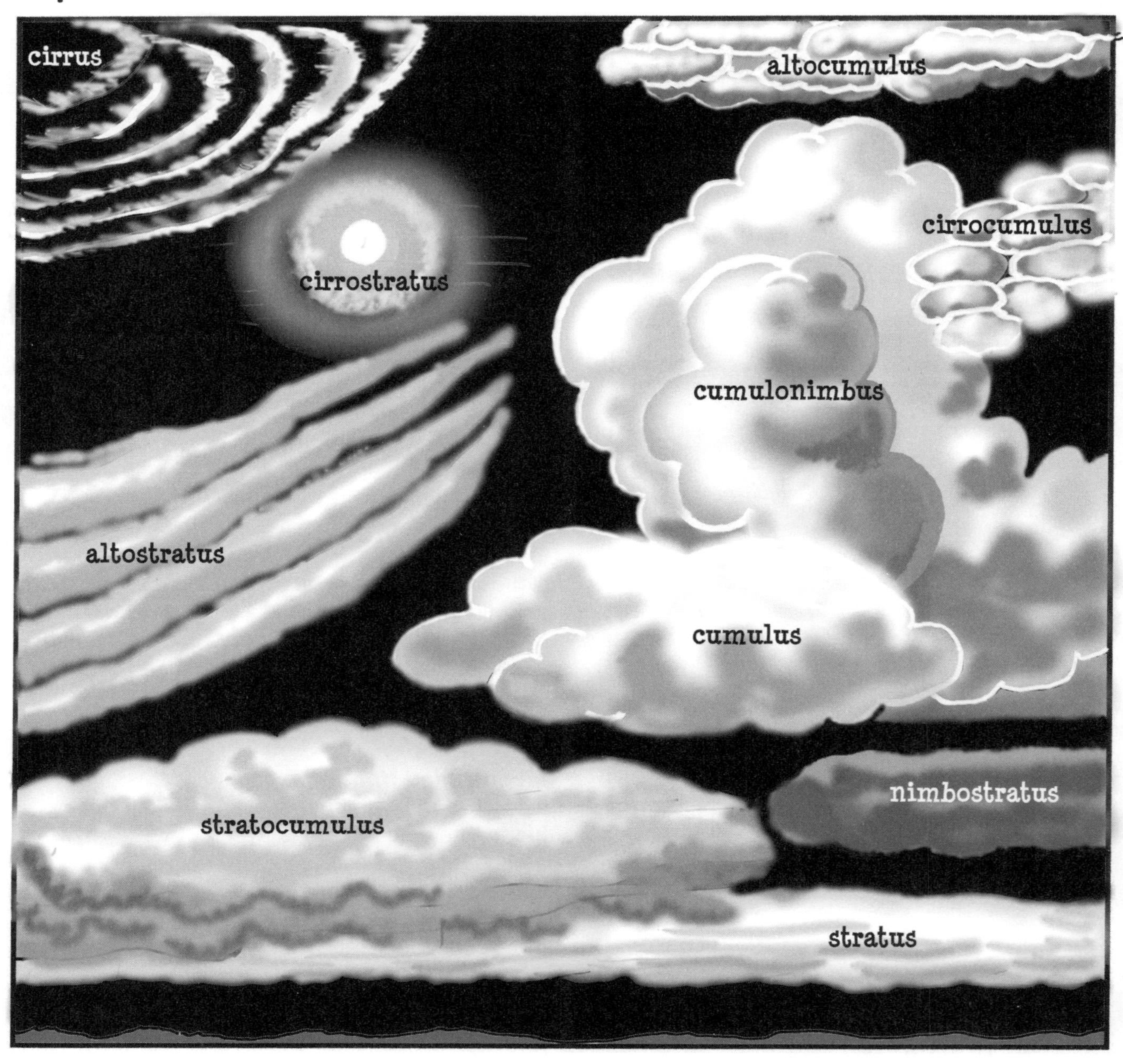

Cumulus	turret-shaped tops, flat bottoms	fair weather
Cumulonimbus	thunderheads (large, dark cumulus)	thunderstorm
Stratus	smooth layers of low clouds	chance of drizzle or snow
Stratocumulus	piles of clouds in layers	chance of drizzle or snow
Nimbostratus	smooth layers of dark gray clouds	continuous precipitation
Altostratus	thick sheets of gray or blue clouds	rain or snow
Altocumulus	piles of clouds in waves	rain or snow
Cirrus	feather-like clouds (made of ice crystals)	fair weather
Cirrostratus	thin sheets of clouds (causes halo around the Sun or Moon)	rain or snow (within 24 hours)
Cirrocumulus	"cottony" clouds in waves	fair weather

Weather Conditions

The weather is bad, and getting worse. Rita Rainey, reporting.

Tiny droplets of water vapor in the clouds will join together to make bigger drops today. Before long, these drops will be falling to the ground as **rain**.

Right now raindrops are falling through a layer of air colder than 3° C. This is freezing rain, or **sleet**.

This weekend, water will freeze on ice pellets in the clouds and make crystals, which will join with other crystals and produce **snow**.

Wild winds and heavy, driving snow have combined to cause the **blizzard** we're experiencing.

Frozen raindrops keep blowing upwards in air currents. Water keeps freezing around each icy stone until it is ready to fall to earth as **hail**.

I'm Fred Frenzy with the Weather Report. he fog is so thick, you can eat it with a spoon.

Water droplets are forming and hovering over the ground causing **fog**. This has formed because the air above ground is being cooled by ground with a cooler temperature.

You can expect **dew**. The ground will get cold enough for water vapor in the air to condense into water droplets. These will form on surfaces such as leaves or ground.

The surface temperature is below freezing this morning, so the water vapor in the air will freeze as it touches the ground and other surfaces, leaving **frost**.

Expect the **thunderstorm** to show itself with lightning and thunder, and possibly heavy rains.

Whirling funnels of air, called **tornadoes**, are expected to form between the bottom of a storm cloud and the ground.

A **hurricane** has developed over the warm, tropical ocean and is hitting with wind strength over 75 mph.

There has been no rain or any form of precipitation for weeks. We are officially in a **drought** period.

This is KZAP News. I'm Lupe Luney reporting from the eye of the storm.

Climate

Climate is the average long-range weather of an area. Average temperatures, amounts and kinds of precipitation, and wind patterns are all a part of the climate of an area. Different parts of the world certainly have different climates. There are many factors that work together to form the climate conditions of a particular place. Topography, winds, ocean currents, altitude, masses of land, and bodies of water all affect climate. A major factor that determines climate is an area's location in relation to the Sun's energy.

Earth's Revolution and Tilt

The Earth's orbit is an oval. Because of this, Earth is closer to the Sun during parts of its orbit than at other times. Earth's revolution, in combination with Earth's tilt, causes the seasons and the climate changes that come with the seasons.

Summer Solstice — On this day, one of Earth's poles is tilted most directly toward the Sun. This occurs in the Northern Hemisphere on June 21 or 22. Areas north of the Arctic Circle have 24 hours of daylight. Summer solstice occurs in the Southern Hemisphere on December 21 or 22. Areas south of the Antarctic Circle have 24 hours of daylight.

Fall Equinox — On this day, Earth's tilt is sideways toward the Sun, so the hours of daylight and darkness are the same in both hemispheres. In the Northern Hemisphere, this occurs on September 22 or 23. In the Southern Hemisphere, this occurs on March 20 or 21.

Winter Solstice — On this day, one of Earth's poles is tilted most directly away from the Sun. In the Northern Hemisphere, this occurs on December 21 or 22. Areas north of the Arctic Circle have 24 hours of darkness. In the Southern Hemisphere, this occurs on June 21 or 22. Areas south of the Antarctic Circle have 24 hours of darkness.

Spring Equinox — On this day, Earth's tilt again is sideways toward the Sun, so the hours of daylight and darkness are the same in both hemispheres. In the Northern Hemisphere, this occurs on March 20 or 21. In the Southern Hemisphere, this occurs on September 22 or 23.

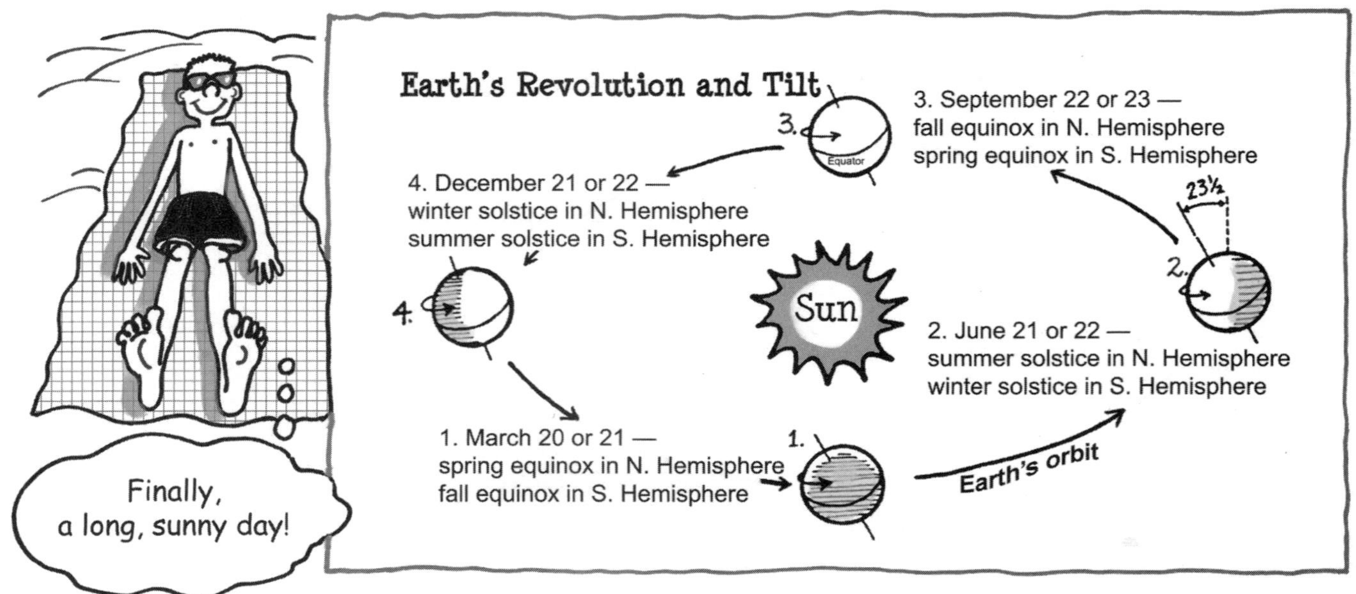

GET SHARP

in

LIFE SCIENCE

Characteristics of Life

What is life?

All living things, or ***organisms***, have some things in common. Every living thing has all of these characteristics.

Every living thing is a simple or complex arrangement of cells.

Cells

Every organism is made of one or more basic units called cells.

Cells are the basic units of every living organism. There are many different kinds of cells, but they all share some basic features.

Plant cells and animal cells differ somewhat. Plant cells have some features that animal cells do not have (vacuoles, cell walls, chloroplasts). Plant cells are usually larger and more oblong than animal cells.

Every living teenager simply desires a cell phone!

Food

All living things need energy for their activities.

They take in some sort of food to give them energy, or they take in some form of energy to make their own food. Every organism needs energy from food in order to carry out the activities of its life.

There are more than 2 million different kinds of living things on Earth.

Water

Every organism needs water to keep living and growing.

Water is an ingredient needed by every living cell. Most living things contain an amazing amount of water in relation to their size. All functions of an organism are dependent upon water.

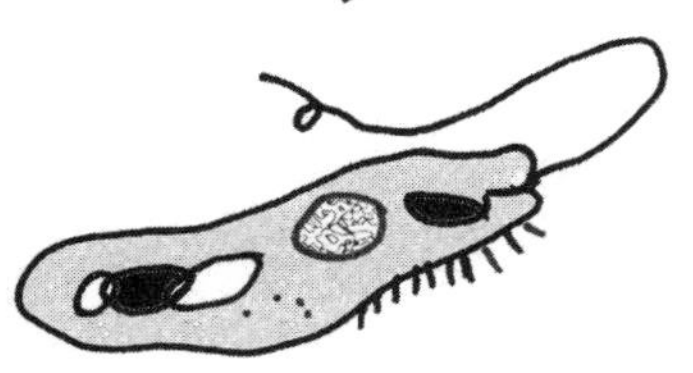

Every living organism has a life span.

Growth & Development

Every organism grows, develops, gets older, and dies.

Many individual organisms also reproduce as a part of their life cycle.

Response

All organisms are sensitive to changes in their environments *(smells, motions, sounds, temperatures, etc.)*. All organisms have ways to respond to the changes they sense.

A turtle might pull into its shell in response to a movement nearby. A plant might change direction in its growth in response to the position of the sunlight.

I've really grown!

Reproduction

Living things are able to make new organisms like themselves. This process is necessary for the continued existence of the species.

Every kind of organism has the ability to reproduce.

Waste

The cells of living things produce waste.

Every organism has some way to give off wastes.

Life Processes

Plasmolysis and osmosis may sound like weird diseases. They're not! They are two of the many processes that take place in living cells. Here is a brief glossary that will help you sort out the different cell processes.

Transport Processes

Transport processes are processes that move life substances around the organism. Most substances will be dissolved in water in order to be transported.

Active Transport

movement of material by cells from areas of lower concentration to areas of higher concentration; the opposite process of diffusion

Diffusion

movement of particles in solution from an area of greater concentration to an area of lower concentration

Osmosis

diffusion of water through a membrane

Plasmolysis

shrinking of the cytoplasm in cells due to loss of water

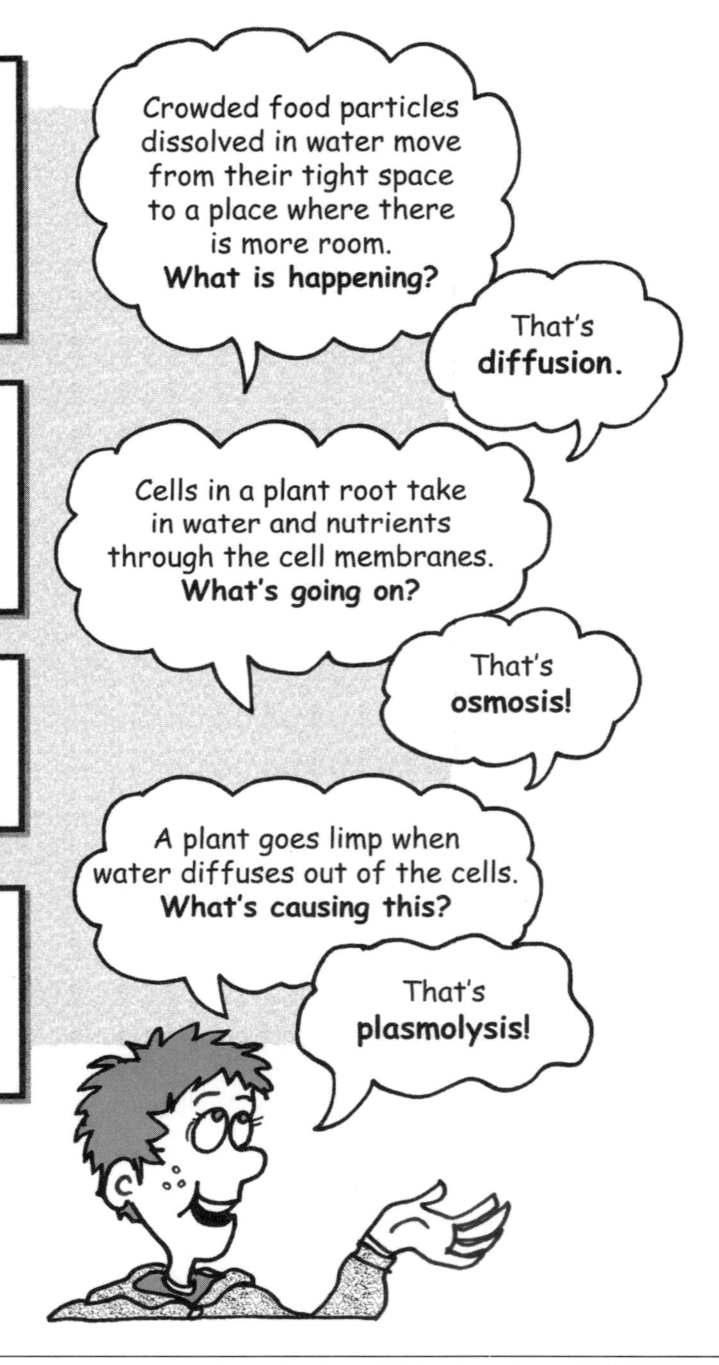

Crowded food particles dissolved in water move from their tight space to a place where there is more room. What is happening?
That's diffusion.
Cells in a plant root take in water and nutrients through the cell membranes. What's going on?
That's osmosis!
A plant goes limp when water diffuses out of the cells. What's causing this?
That's plasmolysis!

Other Processes

Cell Division

the separating of a cell into two identical cells *(identical to the parent cell and identical to each other)*

Reproduction

the process in which organisms create other organisms identical to themselves

Respiration

the process in which cells release energy from food

Mitosis

the division of a cell nucleus that allows the cell to divide

Homeostasis

the tendency of an organism toward balance; a process of adjusting other processes in the cells so the organism is in balance

Metabolism

the sum total of all the chemical processes and changes occurring in an organism

What's happening here?

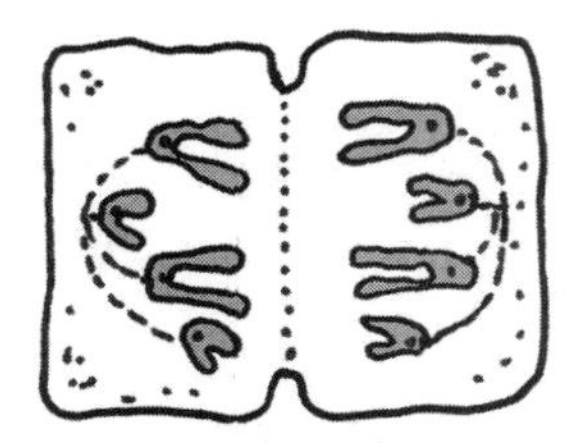

It started with **mitosis**. That led to cell division.

When a swimmer sits on a raft in the wind, her body shivers to warm up her muscles. What's going on here?

That's **homeostasis**.

Life Classification

In order to study living organisms, scientists have divided them into groups that have similar characteristics. One system of classification divides organisms into five large groups, called kingdoms *(moneran, protist, fungus, plant, animal)*. The animal kingdom is then subdivided into six smaller groups (***phylum, class, order, family, genus, species***). The largest groups in the plant kingdom are called ***divisions***. The smallest category for classification in both kingdoms is a ***species***.
Organisms in a species usually reproduce only with other members of the same species.

Get Sharp Tip #13
Taxonomy is the science of classifying living things.

Classification of Life
kingdom
phylum
class
order
family
genus
species

Human

kingdom
Animalia
(animals)

phylum
Chordata
(animals with backbones)

class
Mammalia
(chordates with hair who feed their young with milk)

order
Primates
(mammals with opposing thumbs, large brains)

family
Homonidae
(primates that walk upright)

genus
Homo

species
sapiens

Raccoon

kingdom
Animalia
(animals)

phylum
Chordata
(animals with backbones)

class
Mammalia
(chordates with hair who feed their young with milk)

order
Carnivora
(mammals that eat meat)

family
Procyonidae

genus
Procyon

species
lotor

The scientific name of a species uses the Latin name of the genus and the species. *Procyon lotor* is the scientific name for a raccoon.

Pineapple

kingdom
Plantae
(plants)

subkingdom
Tracheobionta
(vascular plants)

superdivision
Spermatophyta
(seed plants)

division
Magnoliophyta
(flowering plants)

class
Lilipsida
(monocots)

subclass
Zingiberidae

order
Bromeliales

family
Bromeliaceae

genus
Ananas

species
cosmosus

The Five Kingdoms

Monera

The moneran kingdom consists of two phyla of one-celled organisms. ***Bacteria*** mostly absorb food. Some contain chlorophyll. Bacteria can be round, rod-shaped, or spiral in shape. ***Cyanobacteria*** make their own food, contain chlorophyll, and are mostly blue-green in color.

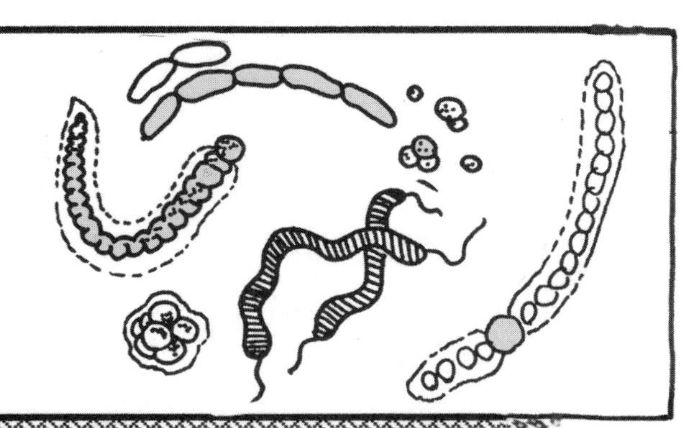

Protista

The protist kingdom has several phyla of mostly one-celled organisms: ***ciliates, euglenoids, golden algae, sporozoans, sarcodines, slime molds,*** and ***flagellates***. Some make their own food, but most take in or absorb food. Most protists move with the help of flagella, pseudopods, or cilia.

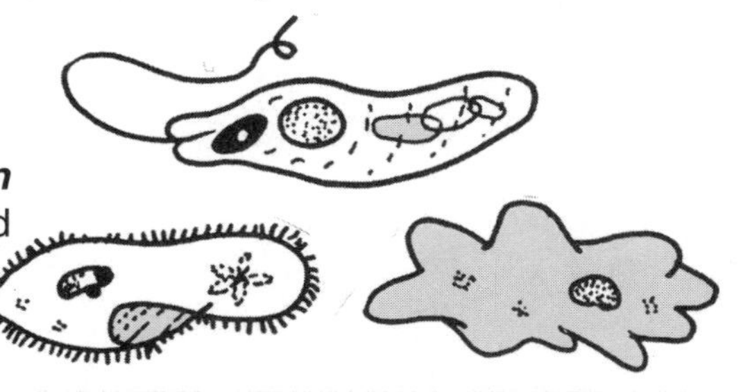

Fungi

The fungus kingdom consists of three phyla. All fungi absorb food. ***Sac fungi*** can be one or many cells. Spores are produced in small sacs. ***Club fungi*** have many cells and produce spores in club-shaped sacs called basidia. ***Sporangium fungi*** have many cells and produce spores in sporangia.

Plantae

The plant kingdom consists of several divisions. Plants are many-celled organisms containing a material called ***cellulose***. Most plants are able to make their own food. Plants are generally anchored in place and do not move from place to place.

Animalia

The animal kingdom consists of several phyla. Animals are many-celled organisms that must find food because they cannot make their own. Most animals are able to move from place to place at some point in their life span.

Plant Classification

The plant kingdom has over 300,000 different species—from simple algae to complex trees and flowering plants. Plants are classified into major divisions. In four of the divisions, the plants are ***nonvascular plants***. These are plants without vessels. Most of them absorb water without the help of roots, stems, or leaves.

Nonvascular Plants

Chlorophyta

(green algae)

This division contains one-celled plants living in colonies, or many-celled green plants living in water or on land. Green algae make their own food. They reproduce by forming zygospores.

Phaeophyta

(brown algae)

This division contains many-celled brown plants. They make their own food and live mostly in salt water.

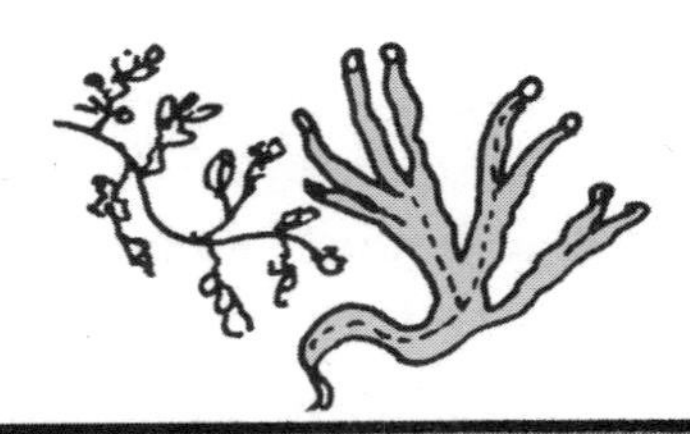

What are **rhizoids** anyway?

Rhodophyta

(red algae)

This division contains many-celled red plants. They make their own food and live mostly in deep salt water.

Bryophyta

(mosses & liverworts)

This division contains many-celled green plants. They make their own food and have a root system of ***rhizoids***. They grow in moist areas on land and reproduce from spores in capsules.

The tallest and longest-living organisms on Earth are plants. The tallest plant is a coast redwood tree standing 363 feet. A bristle-cone pine tree can live 5,000 years.

Rhizoids are wispy, hair-like cells that hold mosses and liverworts to the ground.

Vascular Plants

Vascular plants are plants with vessels. The vessels in plants are a system of tube-like structures that move water with nutrients to all parts of the plant.

Tracheophyta

Club Mosses & Horsetails

Club mosses and horsetails are simple vascular plants. They have many cells and live on land. They are green and make their own food. Both reproduce by forming spores in sporangia.

Most plants make their own food, and don't need to find it elsewhere. But Venus's fly-trap is a flesh-eating plant. It cleverly traps insects to eat.

Ferns

Ferns are many-celled vascular plants. They are green and make their own food. Ferns have feathery leaves called fronds. They live on land or in water. They reproduce by forming spores in sporangia.

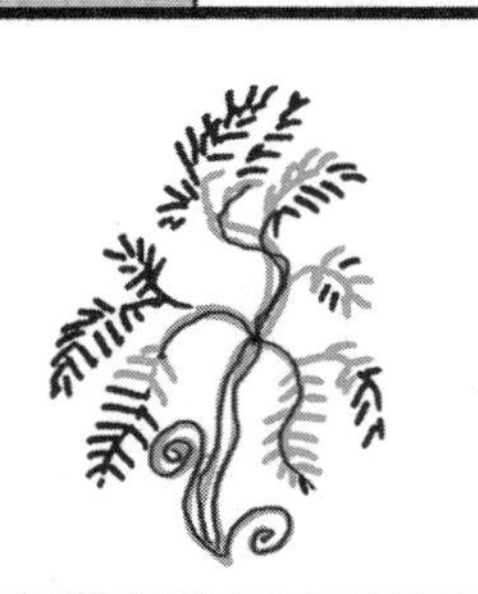

Seed Plants

Seed plants are complex vascular plants with roots, stems, leaves, and seeds. They reproduce by means of seeds that are produced inside a fruit or in cones.

angiosperms
Angiosperms are plants that produce seeds inside a fruit. They are also called ***flowering plants***. Angiosperms have two classes: monocots and dicots.

gymnosperms
Gymnosperms are plants that produce seeds outside of a fruit. Most produce seeds inside cones. (These plants are called ***conifers***.)

Monocots have a single cotyledon (seed leaf) inside their seeds. Monocots have narrow leaves with parallel veins. The petals and sepals grow in multiples of 3.

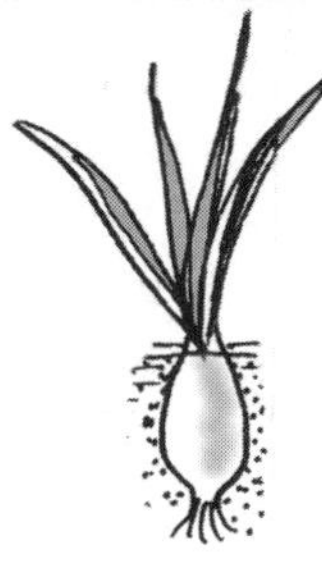

Dicots have two cotyledons inside their seeds. Dicots have broad leaves with branched veins and flower parts in multiples of 4 or 5.

Plant Processes

Plants don't just stand around. They're very busy most of the time doing important work, such as growing, "breathing," and making food. In the process of doing their work, they provide some important supplies for the other living things on the planet.

Photosynthesis

The chlorophyll in green plant cells captures sunlight. The plant uses the energy of the light to combine carbon dioxide with water, producing sugar and oxygen. Sugar is the food the plant will use for its growth and other processes. After it is made, the sugar is stored until it is needed. The oxygen produced is released into the air.

Gas Exchange

Tiny openings in the leaf *(stomata)* allow gases to pass in and out of the leaf by diffusion. Guard cells surrounding each stoma control them, opening them during the day and closing them at night. In gas exchange, carbon dioxide, oxygen, and water vapor pass in and out of the leaves as the plant needs to use or release them.

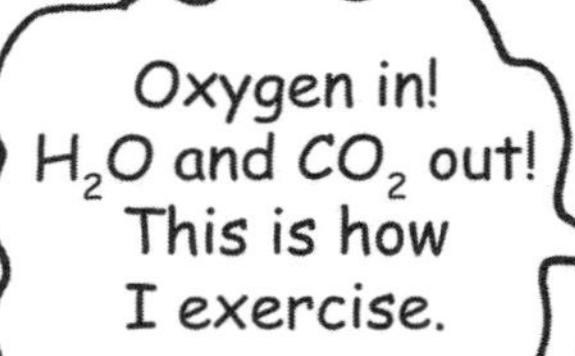

Respiration

Respiration is the opposite of photosynthesis. The plant uses oxygen to break down sugar, releasing energy for its life processes. Water and carbon dioxide are produced as respiration takes place.

Transpiration

Plants lose water vapor through the stomata in the leaves. A plant takes in water through its roots and loses a considerable amount each day through transpiration.

What's The Difference?
Hey, Lily! I get *stimulus* and *response* confused. What's the difference?
I'm glad you asked. A **stimulus** is something happening in the environment that affects the behavior of an organism.
A **response** is a change in the organism's behavior as a result of the stimulus. Plant responses are called **tropisms**.
Do you mean that plants can behave?
Yes, plants behave in response to things that happen around them.
A **stimulus** is something such as: light, absence of light, touch, temperature, changes in air quality, water, and gravity.
How do plants **respond**?
A plant might close up when touched. A flower might wither due to cold temperatures. Roots grow downwards, pulled by gravity. This is called **geotropism**.
A stem can grow towards the light. This is **phototropism**. A flower might bloom as the days grow longer. This is **photoperiodism**.
There are **negative tropisms**, too. For example, roots grow in the opposite direction from light.

Plant Structures

Most plant species have three main organs: roots, stems, and leaves. Each organ serves a particular function for the plant.

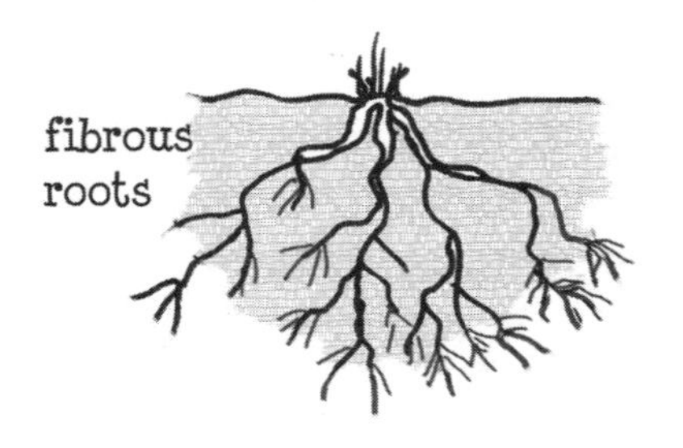

Roots

Roots are the plant's anchors. They hold the plant firmly in the ground so it can't be moved by wind or water. The cells and tissues in roots absorb water and nutrients from the soil.

Taproots – A taproot system is found in plants such as carrots, radishes, beets, or dandelions. These plants have a long, thick, main root that stores food.

Fibrous Roots – Fibrous root systems are made up of branches getting ever smaller as they reach out beneath the plant. Fibrous roots, such as in grass and trees, keep soil in place.

Stems

The stem is a plant's support system, holding the leaves upright. Tissues in the stems also transport food and water around the plant.

Herbaceous stems are green and soft. Most plants with herbaceous stems are ***annuals*** (plants that grow, reproduce, and die all in one year). Beans, petunias, and zucchini plants have herbaceous stems.

Woody stems are rigid and hard, with a covering like bark. Most plants with woody stems are ***perennials*** (plants that do not die after one growing season). Trees and lilac bushes have woody stems.

Leaves

Leaves are the plant's food factories. Leaves trap the sunlight for use in photosynthesis to make food. Other important plant processes, such as respiration, transpiration, and gas exchange, also take place in the leaves.

The **blade** of the leaf is the part the traps the sunlight.

The **petiole** is the leaf stalk that attaches the leaf to the plant stem.

The **epidermis** of a leaf is the thin layer of cells that covers the outer surfaces of the leaf, protecting the inner leaf parts.

Stomata are tiny openings in the epidermis that allow air and water to move in and out of the leaf. Each stoma is controlled by ***guard cells*** that surround the opening and cause it to open and close.

Plant Tissues

Xylem is made up of vessels that move water and nutrients around the plant.

Phloem has cells like tubes. They transport food from the leaves to other parts of the plant.

Cambium is a tissue that makes plant stems grow thicker. It makes new phloem and xylem cells.

Plant Reproduction

Reproduction is part of the life cycle of every plant. It is the way plants create new organisms identical to themselves. The methods of plant reproduction differ, depending upon the simplicity or complexity of the plant. Some methods of reproduction are ***sexual*** (involving material from two different parent cells). Others are ***asexual*** (involving only one parent cell).

Reproduction Without Seeds

Asexual reproduction is found in some of the simplest plants. Only one parent produces an offspring. In some algae, pieces of a parent plant break off and grow into new individual plants. In other algae, reproduction occurs when a cell from the parent plant divides, forming beginning cells for a new plant. In some plants, new offspring can sprout up from underground roots called ***rhizomes*** or ***runners***.

Conjugation is a method of reproduction that results in a zygospore. In some simple plants, cytoplasm from the cell of one plant moves into a cell of another plant. Material from the nuclei of two cells combine (fuse) and form a new cell called a zygospore. This zygospore eventually develops into a new plant.

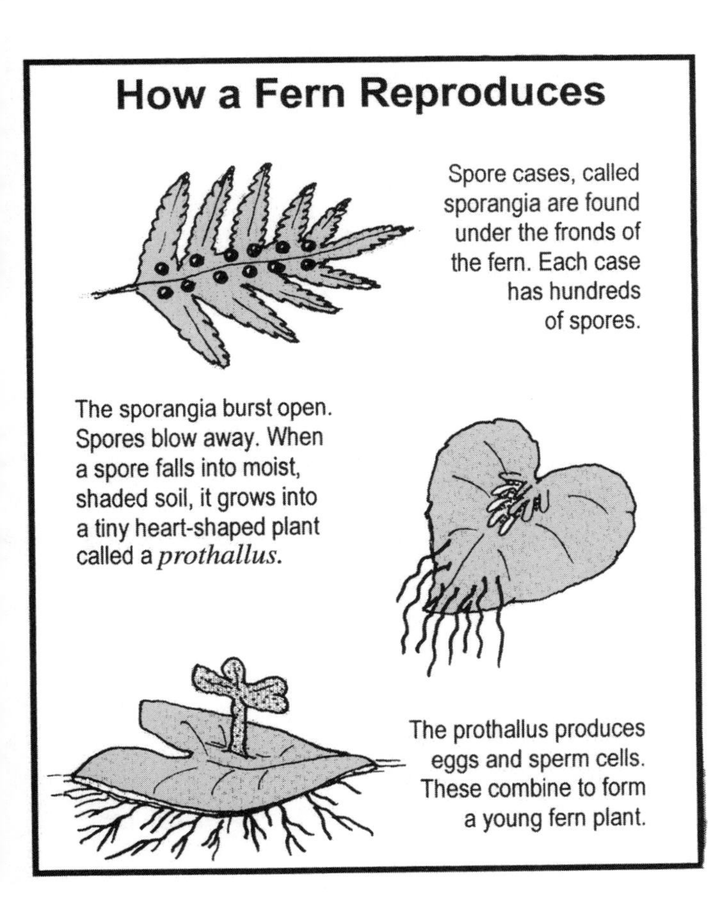

How a Fern Reproduces
Spore cases, called sporangia are found under the fronds of the fern. Each case has hundreds of spores.
The sporangia burst open. Spores blow away. When a spore falls into moist, shaded soil, it grows into a tiny heart-shaped plant called a prothallus.
The prothallus produces eggs and sperm cells. These combine to form a young fern plant.

Reproduction with spores occurs in many seedless plants such as mosses, horsetails, and ferns. A parent plant produces thousands of spores from sex cells called ***gametes***. In ferns, when female cells (eggs) combine with male cells from spores (sperm), a ***zygote*** is formed. It divides and grows into a new fern plant.

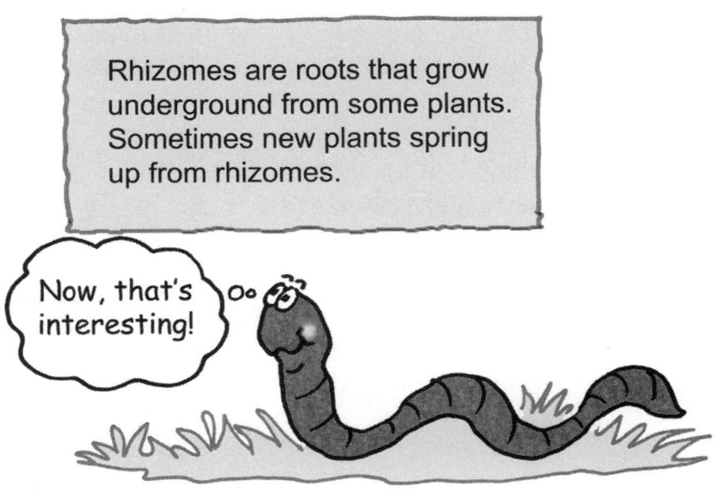

Rhizomes are roots that grow underground from some plants. Sometimes new plants spring up from rhizomes.
Now, that's interesting!

Reproduction in Seed Plants

Many plants reproduce through ***seeds***. A seed is a fertilized plant egg. Seed plants have female and male reproductive organs. These organs are found in flowers (in angiosperms) or in cones (in most gymnosperms).

Here's how reproduction works in seed plants:

- **Eggs**, called ***ovules***, form inside female plant cells.
- **Sperm** form in male cells called ***pollen grains***.
- **Pollination** occurs when a pollen grain transfers to an ovule.
- **Fertilization** occurs after pollination when a sperm cell joins with one of the eggs inside the ovule.
- A **zygote** is the fertilized egg that results from fertilization.
- A **seed** forms from the fertilized ovule.
- An **embryo** (young plant) begins to grow within the seed. The hard covering of the seed protects the young plant as it grows.
- **Dormancy** is a resting period that occurs for many seeds after fertilization. During the dormant period, the embryo does not grow.
- **Germination** is the process where the seed begins to grow from an embryo into a young plant called a ***seedling***. Germination occurs only when temperature, moisture, oxygen, and soil conditions are just right.

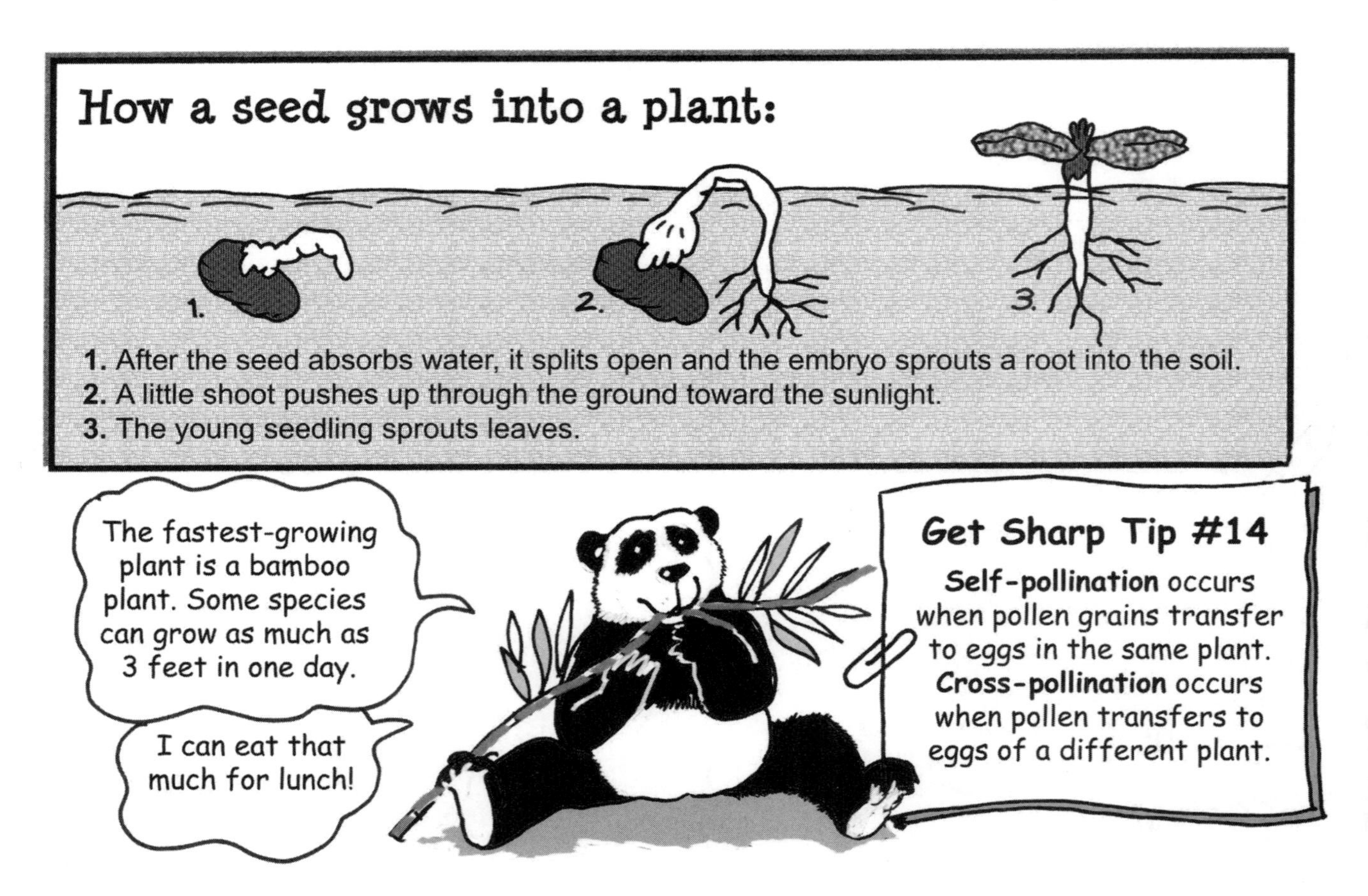

How a seed grows into a plant:
1.
2.
3.
1. After the seed absorbs water, it splits open and the embryo sprouts a root into the soil.
2. A little shoot pushes up through the ground toward the sunlight.
3. The young seedling sprouts leaves.
The fastest-growing plant is a bamboo plant. Some species can grow as much as 3 feet in one day.
I can eat that much for lunch!
Get Sharp Tip #14
Self-pollination occurs when pollen grains transfer to eggs in the same plant. Cross-pollination occurs when pollen transfers to eggs of a different plant.

Reproduction in Flowering Plants

A flower is the reproductive part of an angiosperm. The male part of the flower is the ***stamen***. It has an ***anther*** where pollen is formed, held up by a stalk called a ***filament***. The female part is the ***pistil***. It has an ***ovary*** where eggs are produced inside ovules, a ***style*** (stalk), and a ***stigma***, which is the sticky top of the style. The brightly colored *petals* of the flower attract pollen-carrying insects and other animals.

Pollination occurs when pollen grains from the anther stick on the stigma. A pollen grain sends a tube down the style into an ovule within the ovary. Then a sperm, or male sex cell, travels down the pollen tube and joins with an egg. The result is a fertilized egg, or zygote, which grows into an embryo.

The ovule around the zygote develops into a seed. The ovary around the seeds grows into a fruit. This offers protection to the seeds.

When the fruit dries up and withers, the seeds fall out onto the ground and grow into new plants.

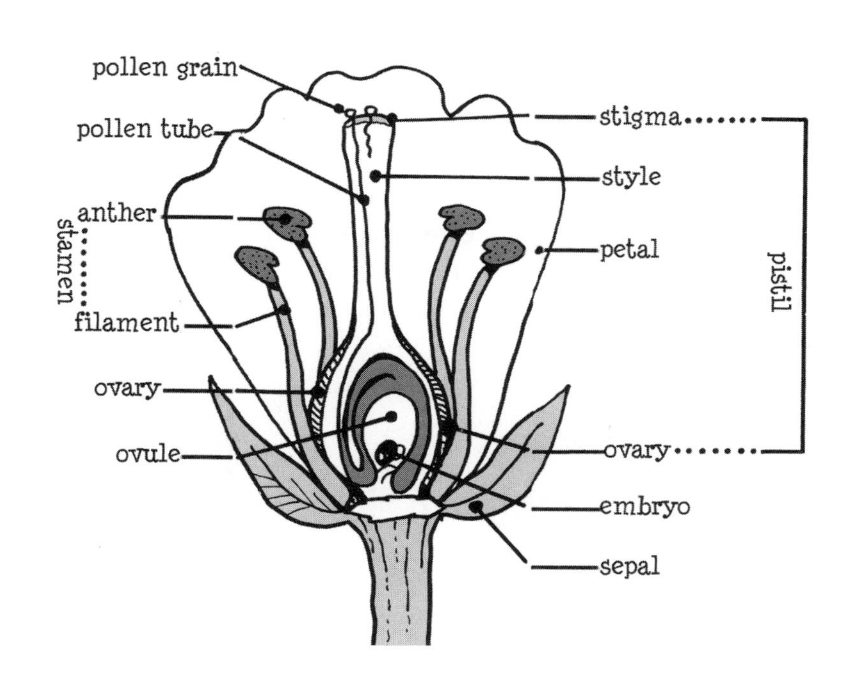

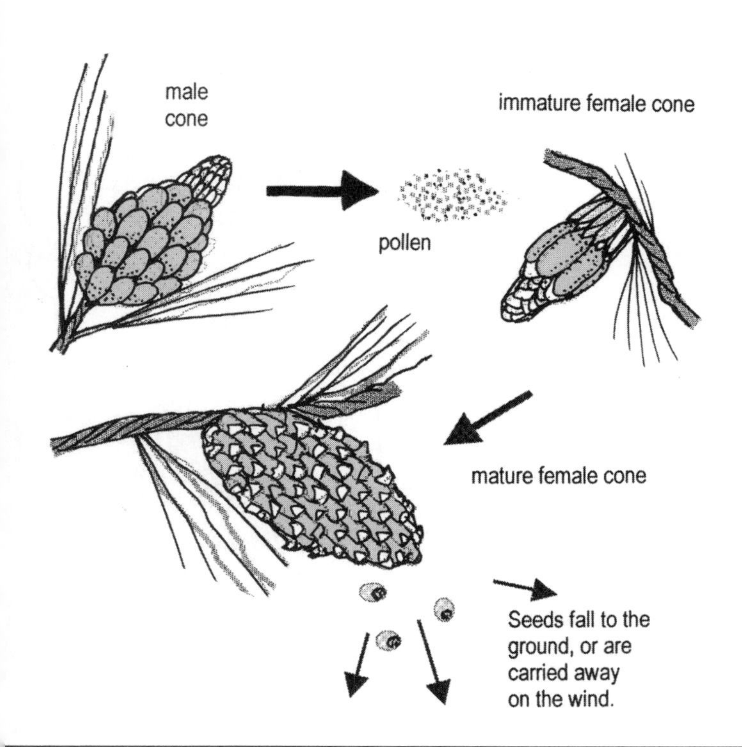

Reproduction in Conifers

In a conifer, male cones produce large amounts of pollen. The pollen is released by the cones and carried by wind to female cones. Pollen grains fall on the female cone.

Each grain sends a tube down into an ovule within the female cone. Sperm then travels down the tube and fertilizes an egg in the ovule. A zygote, or fertilized egg, is formed.

The zygote grows into an embryo and the ovule develops into a seed. Later, the seeds fall from the female cones. Some of the seeds germinate and grow into new plants.

Animal Classification

There are more than a million different animal species, classified into more than twenty phyla. Animals within a phylum share common characteristics. Here are some of the largest phyla.

Sponges pump water through their bodies like a filter. Some can filter their body weight in less than a minute.

Porifera

This phylum contains animals known as sponges. Each organism is a thick sack of cells that form pores, chambers, or canals. They live in water and attach themselves to one place where they stay for life.

Mollusca

Mollusks have soft bodies with hard coverings. They live on land or in water. Many of them have a thick, muscular foot for movement. Some live attached to surfaces such as rocks.
Examples: octopuses, snails, slugs, squids, and clams.

A jellyfish is 95% water.

Coelenterata

Coelenterates have a central cavity with a mouth. They live in water. Most have tentacles. Some stay attached to one place.
Examples: sea anemones, jellyfish, and coral.

There are over 80,000 different kinds of snails. My pet snail's scientific name is *Melanoides tubercularia.*

Would you put a sea cucumber in a salad?

Echinodermata

Echinoderms have radial symmetry and a tough outer covering with spines. They all live in salt water and have a water vascular system for movement. They have tube feet which attach to objects and help them move.

Examples: starfish, sea urchins, sea cucumbers, and sea lilies.

Some starfish can split their bodies in half, then grow enough new legs on both halves to make two new, whole starfish!

Platyhelminthes

The common name is flatworms. These are flat-bodied worms that live as parasites or move freely in the water.

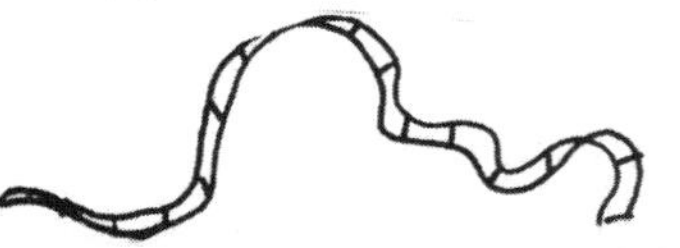

There are more than 15,000 species of nemotodes in the world.

Nemotoda

The common name is roundworms. These are round-bodied worms that live as parasites or move freely in water or on land.

Can you imagine that many worms?

Annelida

The common name is segmented worms. These worms have bodies divided into segments with bristles. They live on land or in water. They are not parasitic.

Arthropoda

Arthropods are animals with jointed limbs, and most have a hard, plated body covering. This is the largest animal phylum.

These are the major classes in the phylum.

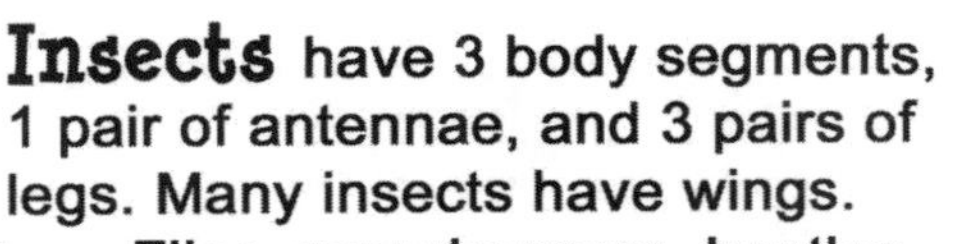

Insects have 3 body segments, 1 pair of antennae, and 3 pairs of legs. Many insects have wings. Flies, grasshoppers, beetles, bees, and butterflies are insects.

Millipedes have segmented bodies with 2 pairs of legs on each segment. They often roll into a ball when threatened.

Arachnids (like spiders, scorpions, ticks, and mites) have an exoskeleton, segmented bodies, 2 body regions, 4 pairs of jointed legs, and no antennae. They have 2 pairs of jointed structures near the mouth used to hold and chew food.

Centipedes have segmented bodies and 2 antennae on the head. There is a pair of legs on each body segment. The mouth has parts that bite and chew.

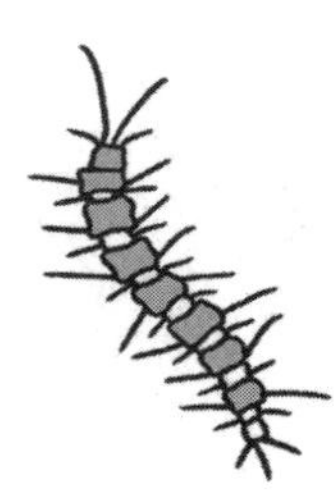

Crustaceans have segmented bodies with two regions. The head and thorax are joined; the abdomen is separate. They have claw-like legs at the end of their bodies, 2 pair of antennae, and sharp jaws for biting and chewing. Lobsters, shrimp, crayfish, and crabs are crustaceans.

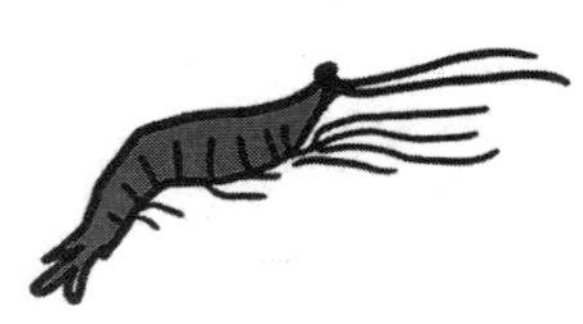

Are spiders fast?

A spider can scurry along at over 1 mile per hour. That's faster than a turtle can move.

Do centipedes have 100 legs?

***Centi* means 100. But not all centipedes have 100 legs.**

Is a seahorse a swimming horse?

A seahorse is not really a horse. It is a fish.

Is a whale a fish, too?

The largest animal in the world, the blue whale, is a mammal. A blue whale can weigh up to 170 tons.

Chordata

Chordates (or vertebrates) have internal skeletons made of bones or cartilage. They also have a central nervous system and specialized body systems for digestion, circulation, and respiration. Some members of this phylum are cold-blooded, and others are warm-blooded animals.

These are the major classes in the phylum:

Amphibians are cold-blooded animals. Their skin is moist with no scales. They breathe air and live on land or water. They have 3-chambered hearts.

Reptiles are cold-blooded animals that breathe air and live mostly on land. Their bodies are scale-covered. They have 3-chambered hearts.

Fish are cold-blooded animals that live in water and breathe with gills. The have 2-chambered hearts, scaly coverings, and skeletons of bone or cartilage.

Birds are warm-blooded aminals with wings and feather coverings. They have hollow, lightweight bones and hearts with 4 chambers.

Mammals are warm-blooded animals that produce milk to feed their offspring. They are covered with hair or fur, have glands that produce sweat, and have 4-chambered hearts.

Animal Reproduction

In order to survive, a species must produce offspring to replace the members of the species that die. Reproduction is the process that solves this problem. All animals reproduce. Here are some of the different methods for reproduction.

Asexual Reproduction

Budding
It's another me.
Hi, you!

In ***asexual reproduction***, there is only one parent and no combination of material from different sex cells.

Budding – Some animals reproduce by growing a new organism from a ***bud*** on the parent. A bulge of extra tissue grows on the parent animal. When the tissue is fully developed into an identical copy of the parent, the tissue breaks off, forming a separate offspring. Budding is common in some simple invertebrates such as sponges, hydras, and sea anemones.

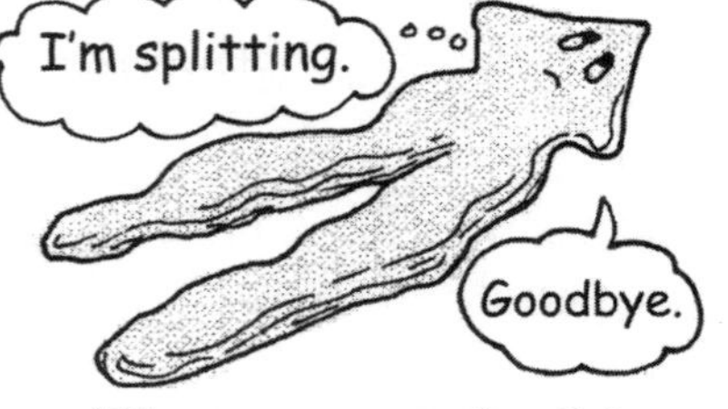
I'm splitting.
Goodbye.
Fragmentation

Fragmentation – Sometimes an animal simply divides into two or more pieces. Each piece grows the missing parts and becomes a whole offspring. Fragmentation is common among many flatworms.

Parthenogenesis – Some insects, such as stick insects, produce offspring that grow from eggs which have not been fertilized.

Sexual Reproduction

In ***sexual reproduction***, material from two different sex cells combines to form a new individual. The joining of a ***sperm*** (male cell) with an ***egg*** (female cell) is called ***fertilization***. This process produces a ***zygote*** (fertilized egg), which then develops into a new individual.

External fertilization takes place outside the animal's body. Often a female releases or deposits eggs, then the male releases sperm which find their way to fertilize the eggs. Most fish, amphibians, and mollusks use this method of fertilization.

Internal fertilization takes place inside the female animal's body. The sperm is deposited in the female's body during mating. In some animals, such as reptiles, birds, and many arthropods, the eggs are deposited and hatched outside the animal's body. In most mammals and some reptiles, the eggs develop inside the female's body. The new animal is born after it is fully developed.

Metamorphosis

There are many species of animals that reproduce sexually but produce offspring that do not have all the same structures as the adult animal. These animals go through a series of changes or stages before taking the shape of the mature adult.

Complete Metamorphosis

In ***complete metamorphosis***, the newborn offspring has little resemblance to the adult animal. Butterflies, bees, ants, flies, and moths all undergo complete metamorphosis.

Butterfly

The butterfly begins with a fertilized egg, or ***zygote***. The zygote hatches into a ***larva***. (A butterfly larva is called a ***caterpillar***.) The larva eats and grows, shedding its skin several times. When its growth is finished, it forms a hard case called a ***pupa***. (A butterfly pupa is called a ***chrysalis***.) The butterfly develops inside the pupa, and hatches from it.

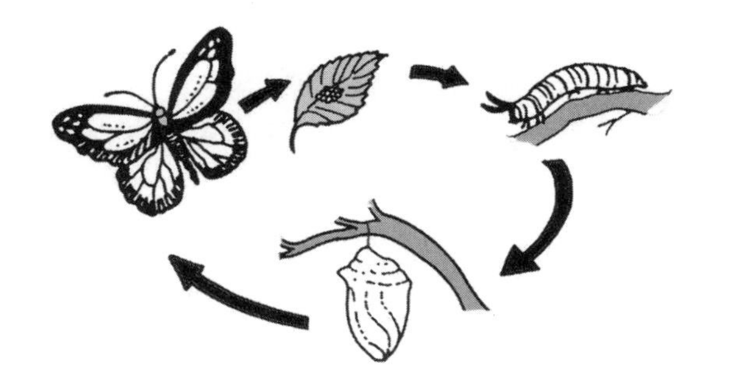

Frog

The frog's eggs develop into ***tadpoles***. A tadpole breathes with gills and eats plants found in the pond. Over time, the tadpole gradually develops complete gills and lungs, and grows legs. As an adult, the frog is able to breath on land, while the tadpole can only survive in the water. The tail is absorbed into the body as the little frog grows.

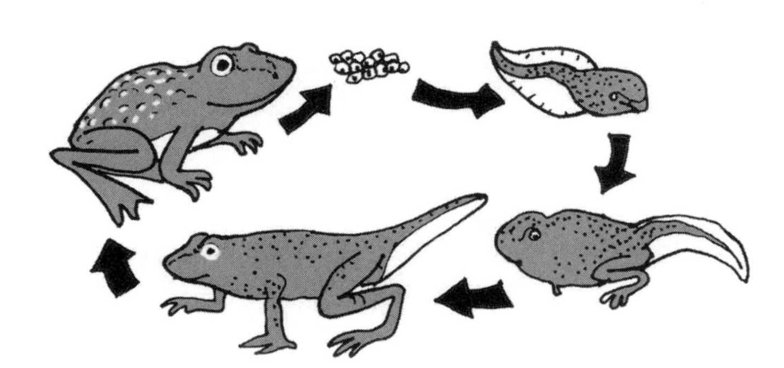

Incomplete Metamorphosis

Incomplete metamorphosis involves three stages of development. The newborn offspring looks like a small adult but does not have all the structures of an adult.

Grasshopper

The tiny grasshopper that hatches from a fertilized egg is called a ***nymph***. It looks like a miniature adult grasshopper, except that it has no wings or reproductive organs. As it grows, it sheds its ***exoskeleton*** (its hard, outer shell) several times and gradually develops wings and reproductive organs.

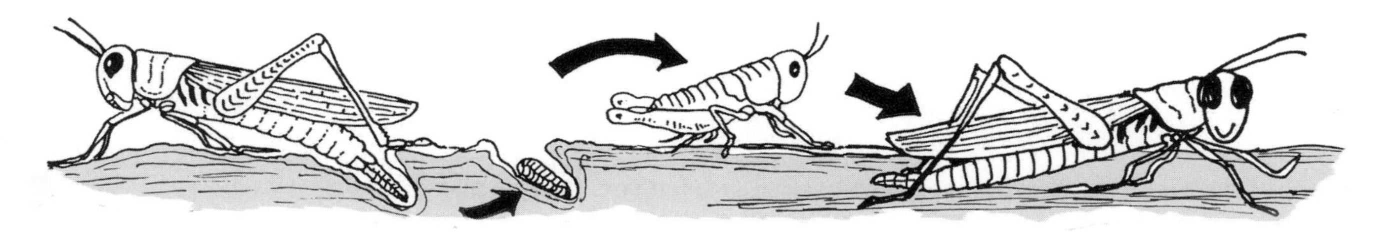

Animal Behavior

Now and then, a darkling beetle will stand on his head. Others beetles roll up into balls. Fiddler crabs change color twice a day. Hognose snakes play dead. Flamingos stand on one leg. African grasshoppers blow bubbles. A daddy frog holds baby tadpoles inside his mouth for weeks. Monarch butterflies fly 4,000 miles to Mexico for a winter vacation. These are just some of the fascinating activities that are part of animal behavior.

adaptation: characteristics or changes that enable an animal survive in its environment

A dog grows thicker fur in cold weather. The woodpecker has a long, sharp bill that is just right for getting to insects living deep inside tree bark.

camouflage: changing body color or appearance for protection

The brown ptarmigan turns white when snow covers the ground.

communication: giving messages, warnings, or other information to each other

Honey bees do a special dance that informs other bees about the supply of nectar. Different dance speeds give different messages about the location and amount of the nectar.

courtship behavior: mating actions

A male dance fly presents a dead insect to a female dance fly. A male pigeon struts around and bows in front of a female pigeon.

defensive behavior: actions that help the animal protect itself against danger

An opossum, threatened by a predator, goes limp and plays dead. A skunk sprays a foul-smelling mist on a predator.

dominance: behavior that shows or keeps an animal's status or power over another animal

A bull seal fights off all other seals that challenge his ownership of a large group of females.

hibernation: a long period of rest or inactivity, usually in winter

A bat eats huge amounts of food in the fall and gets very fat. Then, he sleeps for long periods of time during the winter.

Why would a glass lizard break off its own tail?

That's **defensive** behavior!

The lizard probably spotted a predator, such as a bird. While the bird chases the tail, the lizard gets away safely.

And, don't worry. the lizard will grow a new tail!

instinctive behavior: inborn group of responses to a stimulus; animals do this without being taught

A spider automatically spins a web, though she was never shown how to do it. Birds fly south in the winter, even when not taught to do this.

migration: moving long distances to reproduce, mate, raise young, or find food

The arctic tern travels 22,000 miles round trip to its summer home.

mimicry: copying the appearance or behavior of another animal for protection

The viceroy butterfly looks just like a monarch butterfly, which is convenient, because the monarch tastes terrible to predators such as birds.

movement: getting from one place to another

Rhythmic waves flow through the muscular foot of a snail, pushing it forward or drawing it back. The snail produces a slippery slime, which helps the foot slide along easily.

nurturing behavior: actions of care-taking for offspring

A male king penguin keeps a newborn chick safe and warm under a flap of skin on his feet.

reflex behavior: an automatic reaction to a stimulus, a response that does not involve the brain

A deer, quietly munching leaves in the forest, jerks her head up when a twig snaps nearby. A barefoot child steps on a sharp object. Immediately, she jerks her foot away from the object.

regeneration: growing a new body part after one has broken off

A crab lost a claw in a fight with a predator. The claw grows back in a few days.

social behavior: animals living together in groups to survive, defend themselves, or find food

Wolves gather together in packs for hunting. They have greater success finding and surrounding prey when there are several wolves.

territoriality: actions designed to protect or defend a certain geographical area

A wolf urinates around the edge of the area where he lives. When an owl wanders into a new area, another owl squawks and flaps its wings furiously.

Animal Behavior Quiz II

What kind of behavior is this? A goose hisses at another animal that comes into his home area.

Territoriality!

What kind of behavior is this? An anglerfish looks just like the rocks on the ocean floor.

It's **mimicry** (which, by the way is also a defensive behavior).

Biomes

Every plant or animal lives in a biome. What is a ***biome***? It's a region with a distinct climate, a dominant type of plant, and specific organisms that are characteristic of the region. There are several different biomes. Each one is home to many organisms.

Fresh Water Biome

Characteristics:	Plant & animal life:
– found in rivers, streams, lakes, ponds, swamps, marshes – rich in plant and animal life living in and near the water	– green algae, pond weed, flowers, cattails, fish, crayfish, snakes, turtles, birds, alligators, frogs, insects

Salt Water Biome

Characteristics:	Plant & animal life:
– found in Earth's oceans and seas – organisms are adapted to salt water	– -algae, phytoplankton, fish, seals, whales, sponges, mollusks, coelenterates, and echinoderms

Tropical Rain Forest Biome

Characteristics:	Plant & animal life:
– found near equator – precipitation over 100 in. annually – consistently hot climate, average 80° F temperature	– tall evergreen trees, fruit trees, vines, leafy plants, birds, monkeys, leopards, amphibians, snakes, insects, frogs

Grassland Biome

Characteristics:	Plant & animal life:
- rich soil - 10–40 in. rain annually - hot, dry climate - herds of grazing animals	- thick (tall) grasses, gazelles, zebras, wildebeests, bison, lions, hyenas, vultures, small burrowing animals

Get Sharp Tip #16

A biome does not have clear boundaries. Biomes overlap and characteristics change gradually between biomes.

Desert Biome

Characteristics:
- hot temperatures, cool nights
- less than 10 in. rain annually
- organisms adapted to limited water
- little plant life

Plant & animal life:
- succulent plants that store water, lizards, snakes, rabbits, mice, insects, birds, camels

Temperate Deciduous Forest

Characteristics:
- found in temperate climates
- 40 in. of precipitation a year
- extensive deciduous forests
- many forests have been cleared for farms or homes

Plant & animal life:
- mixed deciduous trees, small mammals, deer, birds, rodents, foxes, insects, wildflowers

Taiga Biome

Characteristics:
- found in cold climates in Northern coniferous forests
- long, harsh winters
- damp ground, fog

Plant & animal life:
- coniferous trees such as spruce, pine, and fir, animals such as bears, wolves, beavers, small mammals, insects

Tundra Biome

Characteristics:
- found near polar ice caps
- constant low temperatures
- permafrost
- land is wet and swampy in summer, frozen in winter

Plant & animal life:
- mosses, grasses, lichen, birds, polar bears, wolves, caribou, walruses, mosquitoes, penguins

Relationships in the Environment

Ecology is the study of the relationships between living things and their surroundings. The area that surrounds an organism, and all the other organisms within that area, is the organism's ***environment***. The particular place within the environment where an organism lives is called its ***habitat***. All the plants and animals that live together in a habitat are a ***community***.

The number of organisms of one kind in a community is a ***population***. In order to describe a community, it is helpful to understand the number and sizes of the different populations. Within an ecosystem, each species has a particular role, called a ***niche***. The niche of a species includes the way that organism interacts with all the other organisms in the ecosystem.

The word ***ecosystem*** is used to describe a community, its habitat, and all of the relationships within that habitat. Ecosystems can be as small as a puddle or a rotten tree stump, or as large as an ocean or a major desert. Every habitat has a limited amount of resources (space, sunlight, nutrients, soil, food, and water). Within an ecosystem, different species make use of different resources, so a delicate balance is formed that allows all the organisms to survive. When any one factor in the ecosystem changes or disappears, the balance may be disturbed.

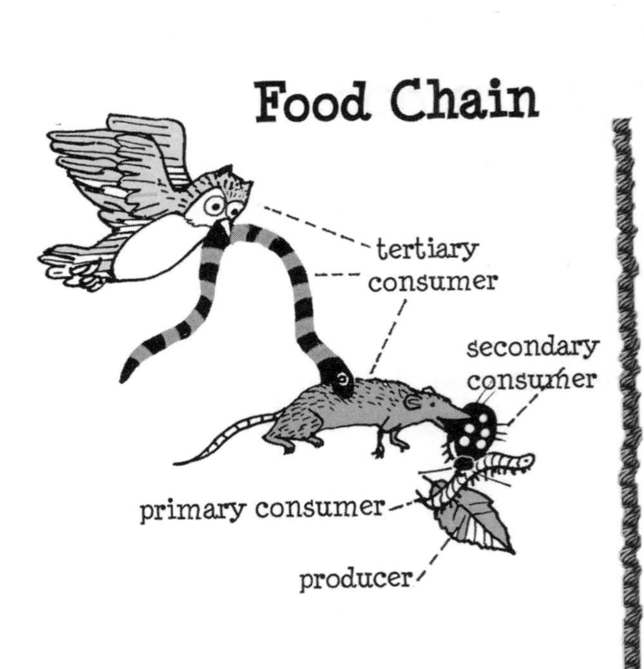

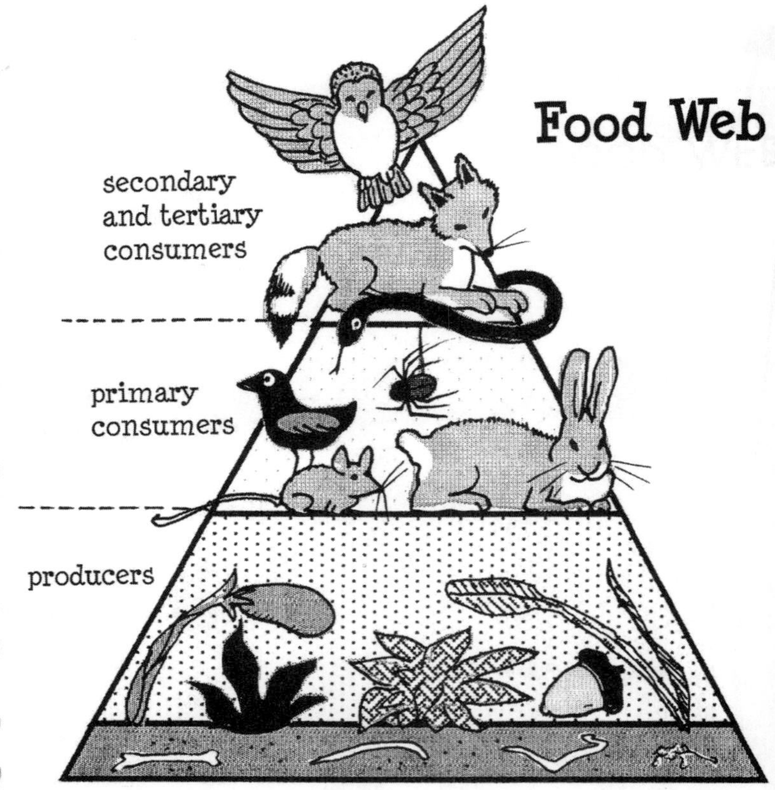

A **food chain** is a series of living things that pass energy along to and through each other. Green plants are the ***producers***. They make their own food. Herbivores (animals that eat green plants) are the ***primary consumers***. Carnivores (meat-eaters) that eat herbivores are ***secondary consumers***. Carnivores that eat other carnivores are ***tertiary consumers***.

A **food web** is a complex network of food chains. Any shortage of one organism in a food web can cause major disturbance in the whole system. If one of the plants doesn't grow well because of drought, if one of the animal species is affected by disease, or if a species is killed, the food supply will be interrupted for all the species above it on the web.

The organisms within an ecosystem interact with each other in a complicated series of relationships. These are some of the major relationships going on in an ecosystem.

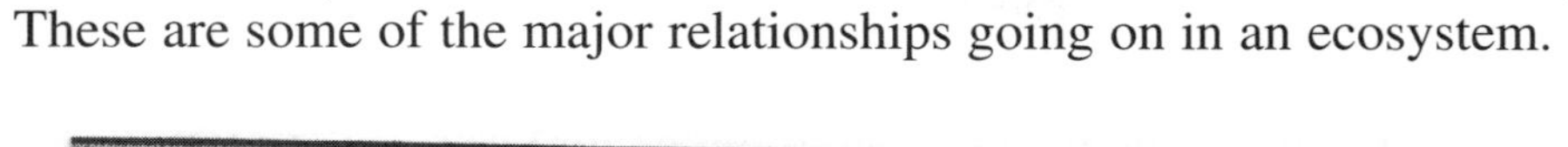

This tick feeds on the blood of other animals. It can cause sickness in the host animal. The tick is a ***parasite***, an organism that lives on a host organism, causing harm to the host.

A fungus grows on a rotten log. It is a ***decomposer***—an organism that helps another organism decay.

Get Sharp Tip #17

Symbiosis is a relationship in which two organisms live in close association with each other. Parasitism, commensalism, and mutualism are all forms of symbiosis.

A worm lives in the shell of this hermit crab. He shares the crab's food. The crab is neither helped nor harmed. This relationship is an example of ***commensalism***.

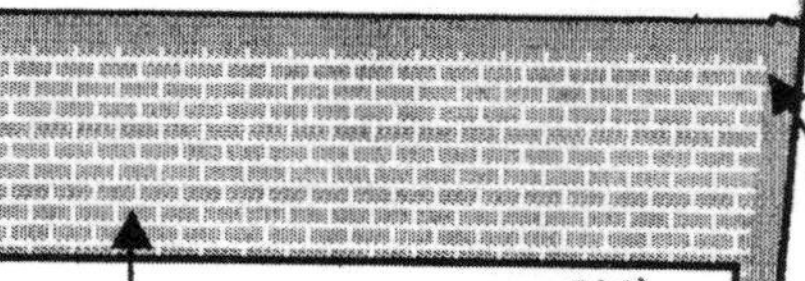

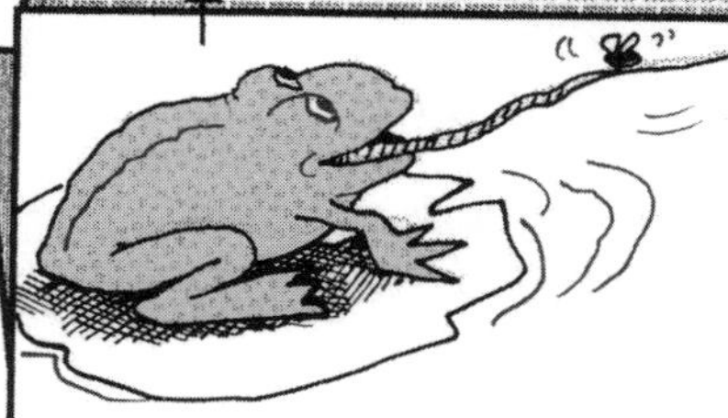

The pond frog catches a fly for dinner on his long tongue. The frog is a ***predator***. The fly is his ***prey***.

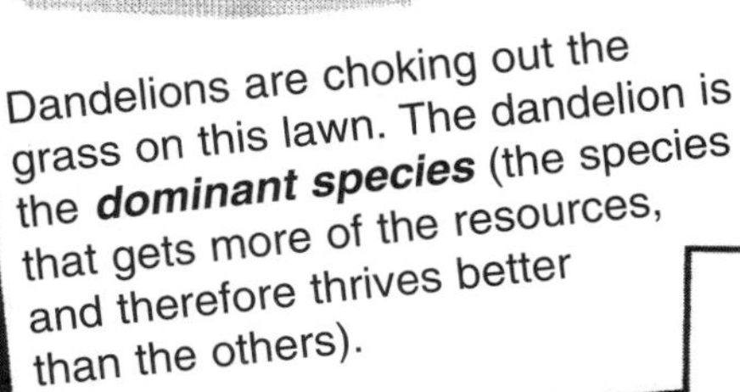

Dandelions are choking out the grass on this lawn. The dandelion is the ***dominant species*** (the species that gets more of the resources, and therefore thrives better than the others).

Ants crawl all over this dead worm, getting food from the worm's body. The ants are ***scavengers***—animals that feed on dead organisms.

The Egyptian plover eats leeches from between the teeth of a crocodile. The plover gets food and the crocodile gets clean teeth. A dependent relationship that benefits both organisms is called ***mutualism***.

Problems in the Environment

The Earth has limited resources, such as water, air, and fuels. The living things on Earth make great demand on those resources. Some living things (such as people) use these resources more heavily than others. Some of the resources are running out. Others are becoming damaged or polluted. ***Pollution*** is the release of harmful substances into the environment. Every day tons of poisonous gases, chemicals, sewage, and garbage are poured into the water, soil, and air. These substances harm or kill organisms and destroy the delicate balance in ecosystems. Other things pollute also. The environment is polluted with noise, unattractive sights, and excess heat.

Sources of Harm to the Environment

FACTORY FUMES
SEWAGE & GARBAGE
PERSONAL SEWAGE & GARBAGE
SPRAYED SUBSTANCES
VEHICLE EXHAUST
BURNING FUELS
PESTICIDES & FERTILIZERS
OIL SPILLS
MINING
ACID RAIN
DROUGHTS, FLOODS & STORMS
SHIPS DUMPING WASTES

Some Terms & Concepts to Know

Acid rain (or snow) is created when sulfur dioxide is released into the air by industries and combines with water vapor in the air. It is harmful to plants, animals, metals, and stone.

Deforestation is the removal of forests, usually to convert land to farms or use them for development. Deforestation can lead to erosion, loss of living species, and fewer plants to use carbon dioxide. This adds to the ***greenhouse effect***.

Erosion is the removal or movement of Earth materials.

Fossil fuels are nonrenewable fuels formed from layers of organisms that have decayed beneath Earth's surface (coal, petroleum, and natural gas).

Global warming is the gradual increase in temperatures on the globe. This may cause excess melting of the polar caps, which could result in serious flooding.

Greenhouse gases such as carbon dioxide are trapped close to the Earth by the atmosphere. As levels of these gases increase, the Earth gets warmer. Greenhouse gases result from the burning of fossil fuels.

Ozone layer is a thick layer about 15 miles up in the atmosphere. Ozone is a form of oxygen that protects the Earth from the Sun's harmful ultraviolet rays. Polluting chemicals called *chlorofluorocarbons* are destroying or "eating a hole" in this protective layer.

Nonrenewable resources are resources that take hundreds to millions of years to form (coal, petroleum, natural gas). The supply is limited.

Renewable resources are resources that can be replaced by nature over a period of time (crops, land, trees, plants, animals).

Sewage is human waste material. Untreated sewage is often dumped into oceans, lakes, or rivers. As it decays, it can be fatal to living things.

Smog is polluted fog. Smog is particularly harmful to the respiratory system. It also blocks sunlight.

Thermal pollution is the raising of the temperature in a river or other body of water when hot water is dumped by industries. The higher temperatures cause growth of organisms that harm fish.

Help for the Troubled Earth

Many methods, programs, changes, and practices are being tried to reduce harm and repair damage to the environment. These are a few of them.

Conservation is a term that applies to any efforts made to protect, replace, or make careful use of resources.

Emissions control is the practice of trying to reduce the amount of harmful emissions released into the environment by automobiles.

Forest management involves reducing waste in cutting and using timber and managing forests to prevent fires, control diseases, reduce erosion, and maintain healthy forests.

Natural pest control is the practice of using harmless, natural substances instead of chemical pesticides to reduce damage to crops.

Recycling is the practice of using things over and over again instead of throwing them away. Paper, aluminum, and glass are easily recycled.

Reforestation is the practice of planting seeds or small trees to replace forests.

Reserves are areas of land that are protected for preserving the habitats of plants and animals.

Soil conservation includes a variety of practices used to reduce erosion and improve the fertility of soil. Some of these methods are the use of mulch, planting cover crops to hold soil, planting trees as windbreaks, contour planting and plowing, and crop rotation.

Use of renewable resources, such as wind power and solar power, slows the use of nonrenewable resources.

Wildlife preservation is the practice of maintaining any living species to protect it from extinction. This usually involves restoring and protecting wildlife habitats.

Some Websites to Explore
Environmental Protection Agency: www.epa.gov
World Wildlife Fund: www.worldwildlife.org
Conservation Kids: www.kidsgowild.com
Conservation International: www.conservation.org
National Resources Conservation Service: www.nres.usda.gov
National and international agreements and organizations, such as the UN Earth Summits, Environmental Protection Agency, World Wildlife Fund, the Sierra Club, and National Resources Conservation Services, have different goals and responsibilities related to preserving Earth's resources.
Recycling is a practice that anyone can do to help the environment. Make it a habit!
Recycle Bin

How the Body Works

The body is a complex machine of different parts. Each part has an important job to do, and the parts work together as systems to move, sense the world, communicate, breathe, reproduce, and get and use energy. If all the body parts are working properly, they function together to keep a healthy, working body.

Cells

The cell is the body's basic unit of life. There are different kinds and shapes of cells in the body, but most body cells are similar in structure. There are more than 50 billion cells in the human body.

Tissues

A ***tissue*** is a group of cells that have a similar shape and function. There are four different kinds of body tissues. Many organs are composed of more than one type of tissue.

Epithelial tissue (skin tissue) forms the outer layer of skin and lines the inner surfaces of the blood vessels and digestive tract.

Connective tissue surrounds organs and connects tissues together. Blood, bone, and cartilage are connective tissues.

Nerve tissue is made up of nerve cells that are able to send signals around the body.

Muscle tissue is made of cells that can contract. Muscle tissue is found in the muscles that attach to the skeletal structures, in the heart, and in other places such as the intestines and stomach.

Organs

An ***organ*** is a group of different tissues that work together to perform a certain function in the body. The heart, for instance, is a group of muscle and nerve tissue. These work together to keep the heart pumping blood around the body.

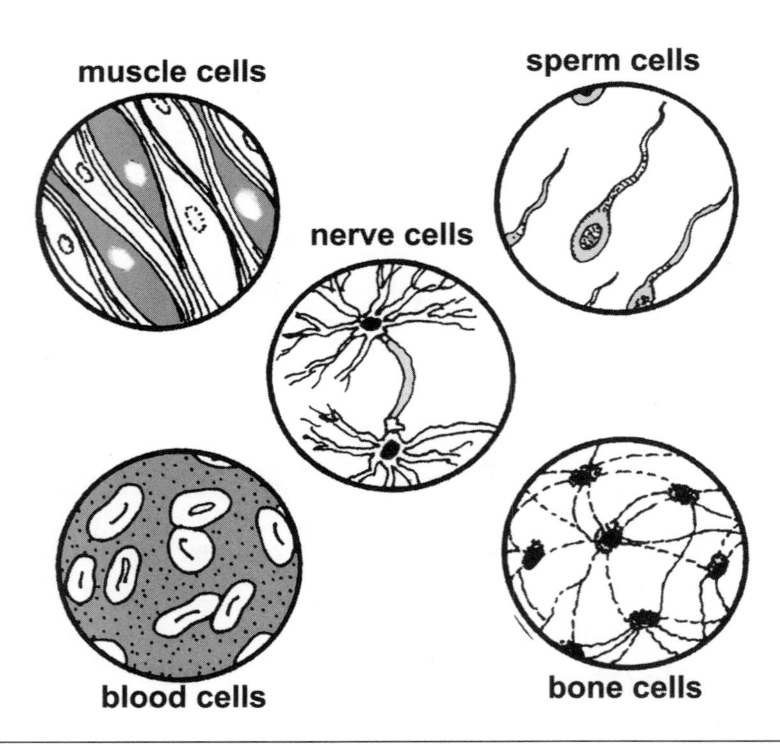

Respiratory System

Cells in the body must have oxygen in order to work. Oxygen is needed for cells to release energy from food. The ***respiratory system*** is a group of organs that takes in oxygen from air, transfers the oxygen to the blood, removes wastes (carbon dioxide) from the blood, and expels the wastes from the body through exhalation.

Breathing

1. As you breathe in, dust and germs are caught by hairs in the nose and mucus in the ***nose and throat***. The air travels through the mouth into the ***pharynx*** (throat), down the ***trachea*** into the left and right ***bronchi*** (structures within the lungs). It continues into the smaller ***bronchioles*** and the tiny ***alveoli*** (air sacs) located at the ends of the bronchioles.
2. The alveoli are surrounded by tiny blood vessels called ***capillaries***. Oxygen passes through the walls of the alveoli into the capillary cells. The capillaries and other blood vessels carry ***oxygen-rich blood*** throughout the body.
3. The blood picks up carbon dioxide wastes from all the body's cells. This is brought back to the capillaries, and transferred through the walls of the alveoli into the lungs. The ***carbon-dioxide laden air*** is breathed out, traveling up from the lungs through the bronchioles, bronchi, trachea, ***pharynx*** (throat), and out the mouth or nose.

The ***larynx*** is the voice box, which contains your vocal cords.

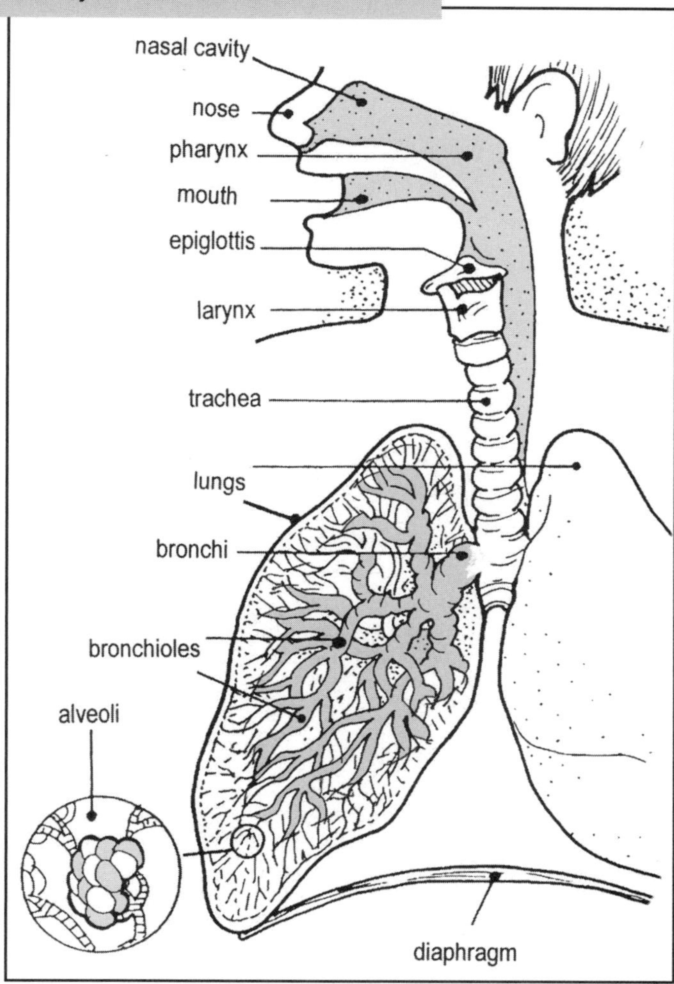

Get Sharp Tip # 18

The epiglottis is the flap at the top of the trachea that closes when you swallow, preventing food from getting into your trachea. This is a good idea because it keeps you from choking!

The ***diaphragm*** is a muscle that helps with breathing. When you inhale, it expands downward, making the chest cavity bigger and allowing the lungs to fill with air. When you exhale, the diaphragm contracts and pushes up, making less room for air and helping to force air out of the body.

Digestive System

Before the food you eat can provide energy for the body's cells, it needs to be changed and processed into a usable form. This process is called ***digestion***. The digestive tract is a long tube that twists and turns its way from the mouth to the anus.

After food is eaten it goes on a 24-hour journey through a 30-foot long tube.

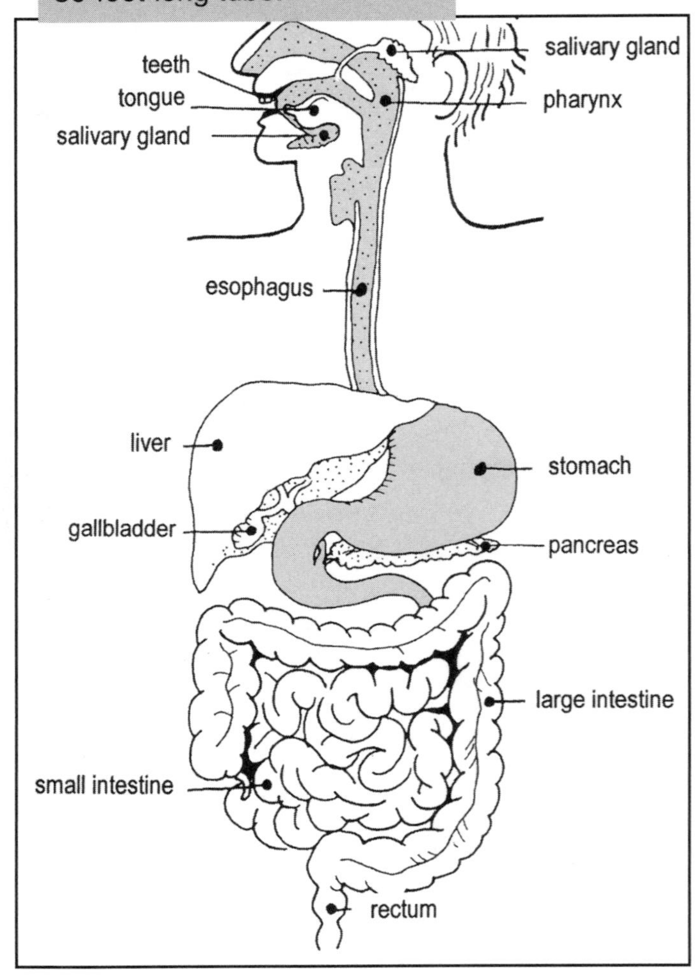

In 1927, doctors examined a woman complaining of intestinal discomfort. It turns out that she had swallowed 2,533 objects, all lodged in her stomach. This included 947 bent pins.

Food's Journey

1. ***Teeth*** bite the food and chew it into tiny pieces. ***Saliva*** made by the ***salivary glands*** helps to dissolve the food with an ***enzyme*** (chemical that changes food).
2. Muscles in the ***tongue*** push the food into the ***pharynx*** (throat). Throat muscles help you swallow the food. As you swallow, the ***epiglottis*** closes over the trachea, so food won't go towards the lungs.
3. The food travels into a long tube called the ***esophagus***, which has muscles that squeeze the food along to the stomach.
4. The ***stomach*** is a powerful muscle. It squeezes and churns food for about four hours, mixing in enzymes which begin digesting proteins. ***Hydrochloric acid*** in the stomach kills any bacteria in the food.
5. The ***liver*** makes a green liquid called ***bile,*** whose job it is to break up fats. Bile is stored in your ***gallbladder***. The ***pancreas*** is a gland that makes digestive enzymes also.
6. When food leaves the stomach, it goes into the first part of the ***small intestine*** (the ***duodenum***) where bile and pancreatic juices mix with the food and digest it. The second half of the small intestine (the ***ilium***), is lined with thousands of tiny finger-like structures called ***villi***. The walls of the villi are one cell. Digested food passes through the walls of the villi into the tiny capillaries within the villi.
7. Undigested food and water move along into the ***large intestine***. The water passes though the walls of the first part of the large intestine (the ***colon***). Solid waste material is stored in the last part of the large intestine (the ***rectum***) until the muscles push it out though the ***anus***.

Waste Removal Systems

Waste products are created as a part of many body processes. These can be poisonous to the cells of the body and cause serious damage or death to body tissues. ***Excretion*** is the process of getting rid of waste products. The body has several ways to get rid of them.

The **digestive system** removes undigested solid wastes and water in the large intestine.

The **lungs** get rid of carbon dioxide and water vapor.

The **skin** is an organ that removes excess water from the body (as *sweat*) to help regulate body temperature. The sweat also contains small amounts of waste mineral material.

Kidneys filter the blood, removing water, mineral, and protein wastes. A liquid called ***urine*** is formed from the filtered wastes. Urine is removed from the kidneys by two long tubes called ***ureters,*** which carry the urine to the ***bladder***, a muscular storage bag. Contractions of the bladder expel urine out of the body through a tube called the ***urethra***.

The **liver** filters harmful chemical substances from the blood.

Oh! It was adrenaline that perked me up during that surprise science quiz.
GLAD ABOUT GLANDS

Endocrine System

Chemical substances called ***hormones*** control many processes in the body. Special tissues called ***glands*** produce the hormones. The hormones are carried throughout the body by the blood. Different glands produce different hormones, each of which has a specific function.

Gland	Gland Location	Hormone Produced	Function of the Gland and Hormone
pituitary	near the brain	11 different hormones	the "master gland'; controls growth; controls many other glands
adrenals	near kidneys	adrenalin	regulates metabolism and body activity in emergencies
pancreas	below & behind stomach	insulin	controls amount of sugar in the blood and storage of sugar in the liver
parathyroid	behind thyroid	parathormone	regulates calcium and phosphorus in the blood and tissues
thyroid	neck	thyroxine	controls metabolism in the body (the rate at which the body uses its food)
ovaries	female pelvic area	estrogen	controls the production of eggs; controls female characteristics
testes	male pelvic area	testosterone	controls the production of sperm cells; controls male characteristics

Skeletal-Muscular System

The human skeleton has more than 200 bones. Along with the muscles, these give strength and shape to the body. They also provide protection for the fragile organs inside the bones, and they are a great place for muscles to attach so the body can move.

Joints occur at places where bones meet. They allow the body to bend and twist. Most joints are padded with a rubbery tissue called ***cartilage***. Bones are held together at joints by strong, flexible bands called ***ligaments***.

There are several kinds of joints. A ***ball and socket joint*** allows the shoulder and hip to swivel. A ***hinge joint*** allows the elbow and knee to bend. A ***sliding joint*** allows the spine to twist and bend. A ***pivot joint*** allows the wrist to twist.

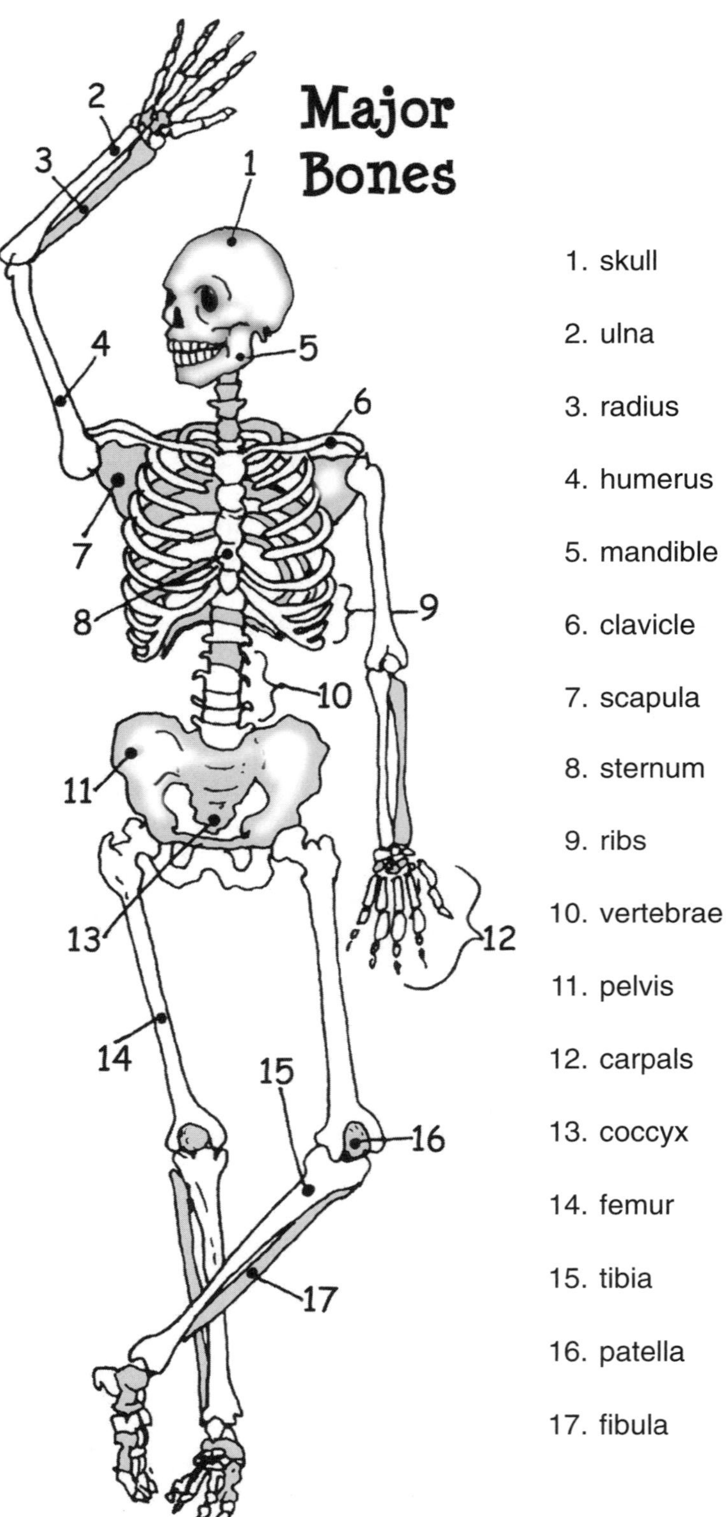

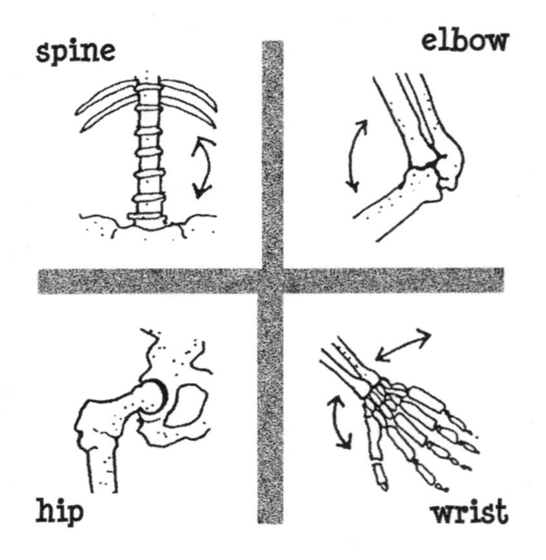

According to the *Guinness Book of Records*, the biggest feet of any living person belonged to Matthew McRory of Pennsylvania. He wore size $28\frac{1}{2}$ shoes! It's no surprise that his mother had to knit his socks!

Muscle Talk
Skeletal muscles generally work in pairs. They pull or relax. They never push.
Look at what happens when my biceps muscle contracts to pull my arm up. It bulges!
When the biceps contracts, the triceps muscle is relaxed.
When I straighten my arm, the biceps relaxes and the triceps contracts to pull the arm down.

Muscle tissue is used to move all parts of your body. It even helps the beating of your heart and the movement of food through the digestive system. ***Voluntary muscles*** move when you choose for them to move. ***Involuntary muscles***, like the heart muscle, muscles in the intestines, or muscles that blink your eyes, move automatically.

The biggest biceps in the world belongs to a man who started building his muscles wrestling pigs. Denis Sester's biceps measures over 30 inches.

Striated muscles are voluntary muscles. These are also known as ***skeletal muscles*** because they move the bones of the skeletal system. They are called striated because they look striped. These muscles are attached to the bones with tendons.

Smooth muscles line the insides of many body organs. Smooth muscles are involuntary.

Cardiac muscle has only one location—that's in your heart. It is an automatic, or involuntary, muscle.

A look inside a bone

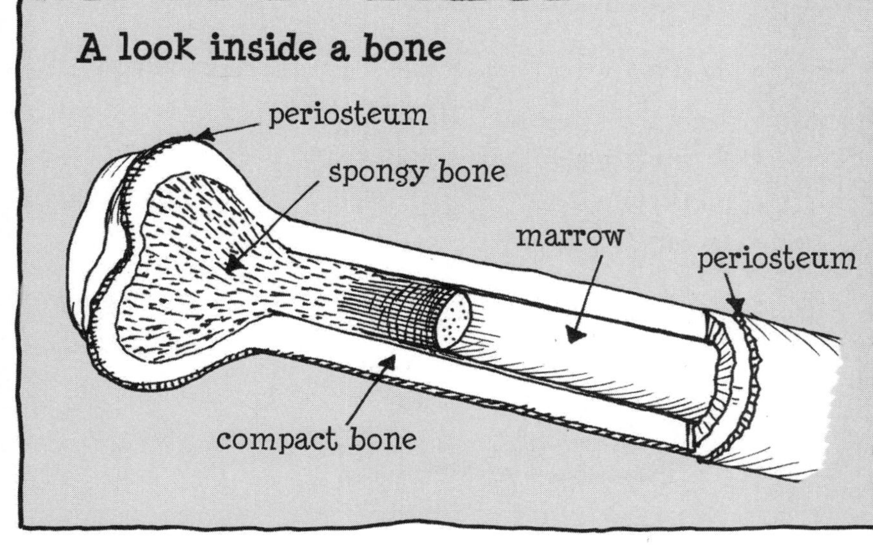

The outer layer of the bone is a very thin, tough layer, called the ***periosteum.*** When a bone breaks, the periosteum cells multiply and grow over the break. Hard ***compact bone*** lies beneath the periosteum. The inner bone is ***spongy bone***. It is light, with lots of holes, but it is very strong. Bone gets its strength and hardness from phosphorus, calcium, and a protein material called ***collagen***. Many bones have a soft, inner tissue called ***marrow***. Red blood cells are made in the ***red marrow.***

Nervous System

The ***nervous system*** is a busy and complex network of nerves. Each nerve is a bundle of nerve fibers that extend from nerve cells or ***neurons***. Most of the bodies of the nerve cells are located in the brain or spinal cord. The ***brain*** and the ***spinal cord*** are the ***central nervous system***. All of the rest of the nerves make up the ***peripheral nervous system*** and carry messages between the central nervous system and the rest of the body. The nerves send signals to and from the brain, and control ***autonomic*** (automatic) processes such as digestion, circulation, and breathing.

Neurons

Neurons* (D)** are cells that carry ***impulses (electric signals). A neuron has a cell body with a ***nucleus*** **(F)** and ***dendrites*** (long branches) **(C),** and ***axons*** (long fibers) **(B).** Some have ***receptors* (A).** Receptors in sensory neurons receive a stimulus and pass it along the axons to the dendrites of connector cells. Impulses travel across ***synapses*** (small spaces between cells) **(E)** until they reach motor neurons.

Sensory neurons carry impulses from receptors in sense organs to the central nervous system; ***motor neurons*** carry impulses away from the central nervous system to muscles and glands; and ***interneurons*** connect sensory neurons and motor neurons.

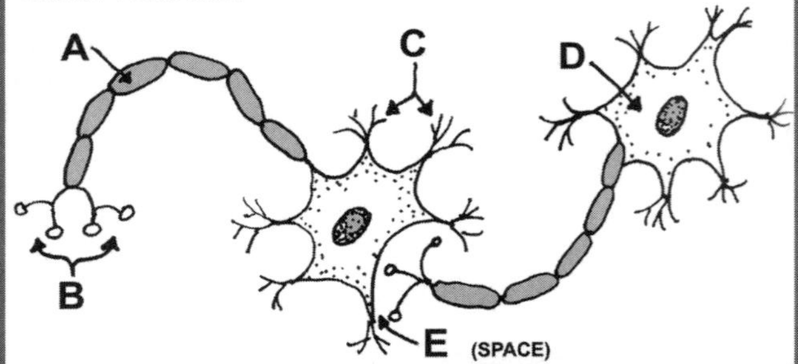

The Brain

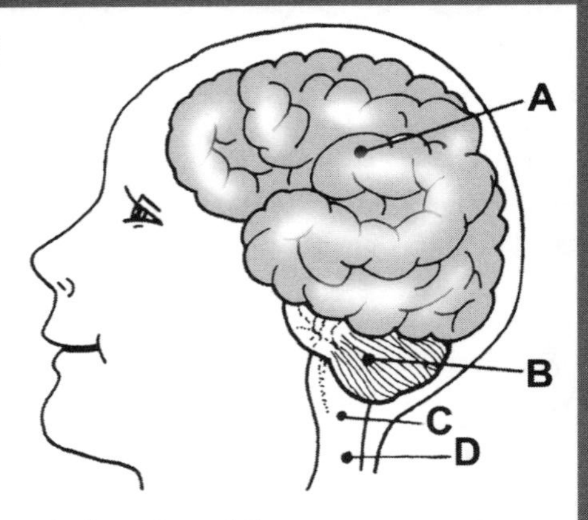

The brain is the control center of the nervous system. The ***cerebrum* (A)** is the largest part. It controls thinking, learning, memory, awareness, and some voluntary movements. It also controls all senses. At the back of the head lies the ***cerebellum* (B)**. It controls muscle activity and balance.

The ***medulla* (C)** is the smallest part of the brain. It lies at the base of the skull. It controls automatic functions such as breathing, the heartbeat, gland secretions, reflexes, and digestive action. The ***spinal cord* (D)** is a thick cord of nerves that runs through the vertebrae. It carries messages between the brain and the rest of the body.

Touch

The skin is a sensory organ that has many nerve endings. It is sensitive to touch, heat, cold, and pain. The top layer, the ***epidermis*** **(A),** keeps out germs. The second layer, the ***dermis*** **(B)**, is thicker. It contains the ***nerve endings*** **(C)**, ***sweat glands*** **(D)**, ***blood vessels*** **(E)**, and ***hair roots*** **(F)**.

Sweat glands help to regulate the body's temperature by passing sweat out of the ***pores*** **(G).** As the sweat evaporates, the body cools.

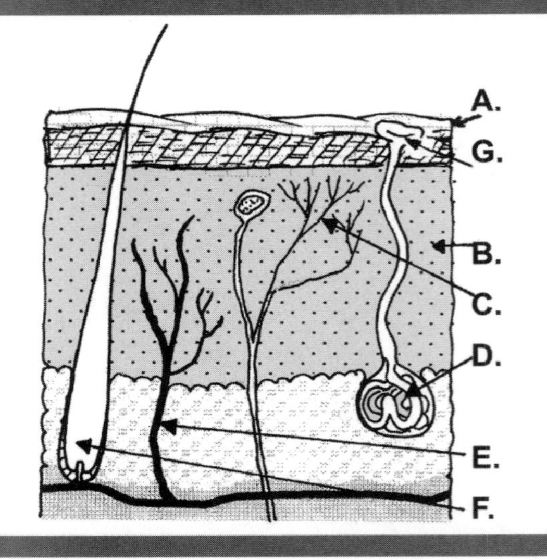

The skin areas on your fingers, lips, and the soles of your feet have more sensory neurons than anywhere else.

Sight

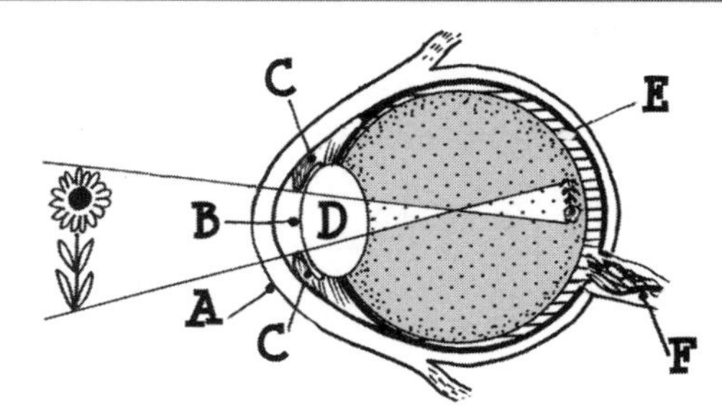

Every image constantly reflects rays of light. Carrying the reflected image, the rays pass through the ***cornea*** **(A)**, the eye's thin protective outer layer, and enter into the eye through the ***pupil*** **(B).** The pupil is a small hole in the center of the eye, surrounded by the ***iris*** **(C)**, a muscle that controls the size of the hole. The light passes through a transparent disc called a ***lens*** **(D)**, which bends the light to focus it on the retina. The image is upside down when it reaches the retina. The ***retina*** **(E)** is the "screen," an area in the back of the eye that contains receptor cells. These cells send impulses through the ***optic nerve*** **(F)** to the brain, which interprets the signals. The brain turns the image right side up.

Hearing

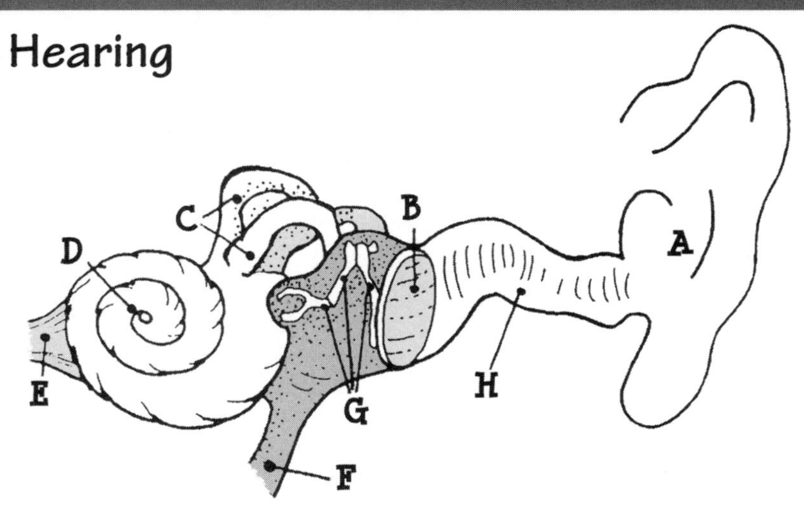

The ear is the sensory organ that allows you to hear. The ***outer ear*** **(A)** catches sounds and directs them into the ear, down a tube called the ***auditory canal*** **(H),** which is lined with hairs and produces wax. The ***eardrum*** **(B)**, a stretched membrane, separates the outer ear from the ***middle ear***. The eardrum vibrates when sound hits it. Three tiny bones in the inner ear, the ***hammer, anvil,*** and ***stirrup*** **(G)**, pick up vibrations and pass them along to the ***inner ear***. The ***cochlea*** **(D)*,*** a coiled liquid-filled tube, picks up vibrations from the middle ear and passes them along to the ***auditory nerve*** **(E)**. This nerve sends sound signals to the brain. The ***semicircular canals*** **(C)** in the inner ear help to maintain balance while the body is in motion. The ***Eustachian tube*** **(F)** leads from the back of the nose to the ear. This is the only way air can get in and out of the middle ear to keep pressure equal on both sides of the eardrum.

Smell

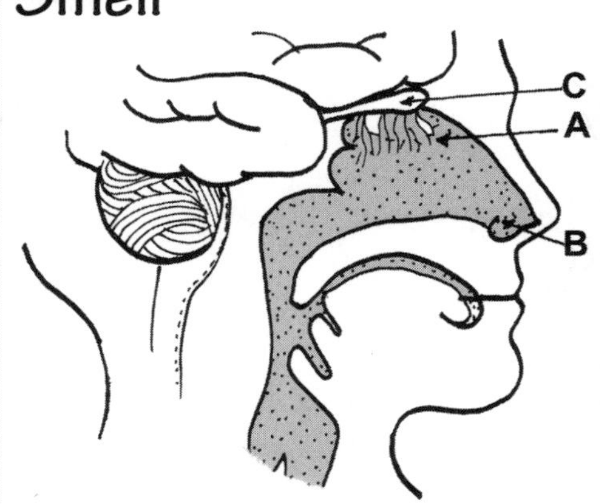

The nose is the organ responsible for the sense of smell. ***Scent-sensitive cells*** **(A)** in the back of the nose respond to scents dissolved in ***mucus*** in the ***nose*** **(B)**. Signals are carried from those cells by other nerve cells to the brain's ***olfactory lobe*** **(C)**. This is the area where smells are recognized.

Taste

Taste is closely related to smell. The sense of smell is actually much stronger than the sense of taste, so tasting relies partly on the sense of smell. ***Taste buds*** at the front, sides, and back of the tongue are groups of receptor cells that are sensitive to chemicals dissolved in the saliva. Buds in different areas of the tongue respond to different tastes.

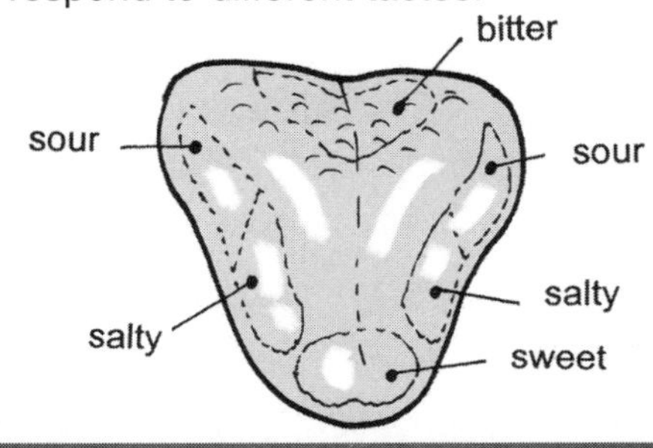

In 1978, a man ate 91 pickled onions in just over 1 minute.

Which of the 10,000 taste buds allow a person to taste a pickled onion?

Circulatory System

The ***circulatory system*** is a vast network of large and small tubes called ***vessels*** that carry blood around the body. The blood transports valuable supplies of oxygen and food to all of the body's cells and carries wastes away from the cells.

Blood

Blood consists of a mixture of different kinds of blood cells floating in a yellow liquid. Blood cells are made inside the marrow of bones.

Red blood cells carry oxygen. As blood passes through the lungs, oxygen combines with a chemical compound called ***hemoglobin*** that is contained in the red blood cells.

White blood cells defend the body by producing antibodies to fight off disease and by engulfing harmful bacteria.

Platelets are fragments of blood cells that stop the bleeding when a vessel is damaged.

Plasma is the liquid that carries blood cells and platelets through the vessels.

Blood Vessels

Blood flows out from the heart through vessels that divide into smaller and smaller vessels, eventually becoming capillaries that surround body tissues and organs. The capillaries join together again to form larger and larger vessels carrying blood back to the heart.

Arteries are the vessels that carry blood away from the heart. They have thick walls because the heart pulses blood through them at a high pressure. The ***aorta***, the main artery leaving the heart, is the body's largest artery. The ***carotid artery*** supplies blood to the brain.

Veins are the vessels that carry blood back to the heart. The walls of veins are thinner because the pressure is lower than in arteries. There are valves in veins to keep the blood from flowing backwards. The ***superior vena cava,*** coming into the heart, is the largest vein in the body.

Capillaries are the tiniest vessels, with walls one cell thick. Substances pass between the capillaries and the cells, delivering food and oxygen, and carrying away wastes.

Blood Types

Everyone's blood falls into one of four groups: type A, B, AB, or O. For a blood donation, the donor and patient must be matched carefully because some types of blood contain antibodies against other blood types.

Blood Type	*Can Receive*	*Can Donate to*
A	**A, O**	**A, AB**
B	**B, O**	**B, AB**
AB	*all types*	**AB**
O	**O**	*all types*

Get Sharp Tip #19

Blood in arteries looks red because it is carrying oxygen combined with hemoglobin.
In veins, blood looks blue because it is loaded with carbon dioxide wastes.

St. Vitas Hospital
I like any type of blood.
That's why I'm visiting the blood bank.

The ***heart*** is the pump that makes the whole circulatory system work. It is a strong muscle that pumps blood through a network of almost 70,000 miles of blood vessels. The heart is divided into four ***chambers***. Each side of the heart has an upper chamber (an ***atrium***) and a lower chamber (a ***ventricle***). As the blood flows through the heart, ***valves*** open and close between each atrium and ventricle, and between the ventricles and the arteries leaving the heart. The valves keep the blood from flowing backwards.

St. Vitas Hospital Newsletter

January 12

Ask Dr. A. Orta

Dear Dr. A. Orta,

What causes a pulse?

Missy Heartfeldt

Dear Missy,

When the heart muscle contracts, arteries pulsate or throb as blood surges through them, pushed by the heart. This pulsation is what you can feel by pressing fingers on the wrist or at the side of the neck.

Dear Dr. A. Orta,

My heart beats faster when cute boys are around. Why?

A Teenager in Peoria

Dear Teen,

An adult normally has a resting heart rate of 70 beats per minute. (This means your heart beats more than 100,000 times a day!) Children have a faster heart beat—about 100 beats per minute. The heart beats even faster than this if you are running, exercising, or falling in love.

Dear Dr. A. Orta,

Why does my heart make a drumming sound?

Drummer in Des Moines

Dear Drummer,

The sound of the heartbeat is caused by the slamming shut of heart valves. The first beat is made by the valves between the atria and the ventricles. The second beat is the closing of the valves between the ventricles and the arteries leaving the heart.

The Heart

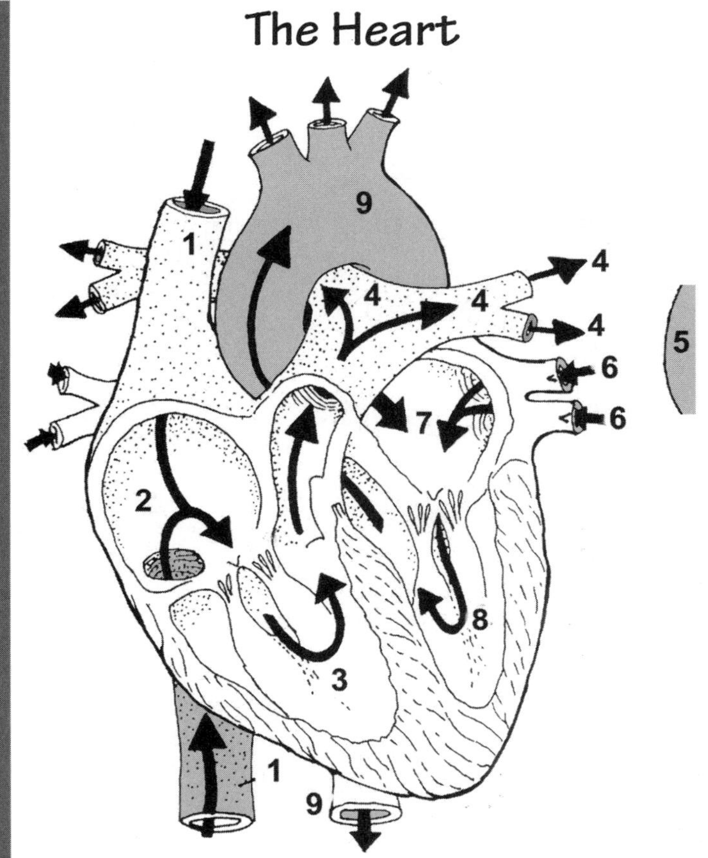

1. The largest vein (the ***superior vena cava***) brings blood that is low in oxygen and high in carbon dioxide into the heart.
2. The blood enters the ***right atrium***.
3. The right atrium contracts and forces blood through a valve into the ***right ventricle***.
4. The right ventricle contracts and forces blood through the ***pulmonary arteries*** toward the lungs.
5. In the lungs, the carbon dioxide is exchanged for oxygen.
6. This oxygen-rich blood flows back from the lungs through the ***pulmonary veins***.
7. The oxygen-rich blood from the lungs enters the ***left atrium***.
8. The left atrium contracts and forces blood through a valve into the ***left ventricle***.
9. From the left ventricle, the heart pumps the blood out through the ***aorta*** into the body.

Reproductive System

Like other mammals, humans reproduce sexually. This means it takes two cells for reproduction to occur: a female egg cell and a male sperm cell. These cells join together in a process called ***fertilization***.

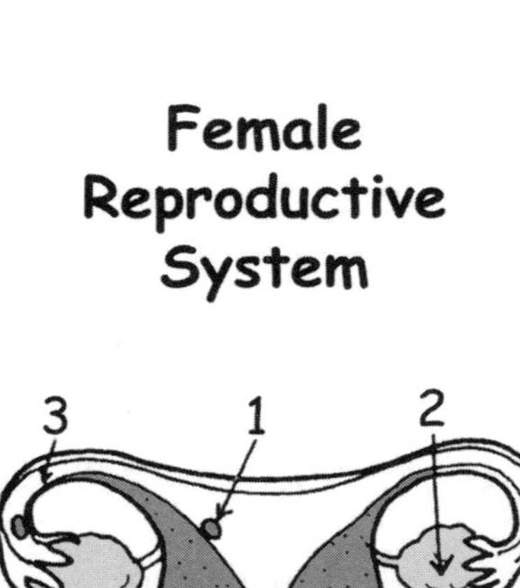

A female releases an ***egg*** **(1)** each month from one of the ***ovaries*** **(2)**. The egg travels down the ***fallopian tube*** **(3)**.

500 million ***sperm cells*** are made each day in the ***testes*** **(7)** of the male. The sperm cells are stored in a coiled tube called the ***epididymis*** **(9)**. The sperm cells mix with a fluid that is made in the ***prostate gland*** **(11)**. The fluid containing the sperm is called ***semen***. The semen is ejected out of the ***penis*** **(8)** through ***sperm ducts*** **(10)** into the ***vagina*** **(6)** of the female.

When reproduction occurs, an egg is fertilized by a sperm that reaches and penetrates it after a successful swim from the vagina through the ***cervix*** (opening) **(5)** and the ***uterus*** **(4)** into the fallopian tube.

When an egg is fertilized, it travels into the ***uterus*** and attaches itself to the uterine wall. The egg divides and grows into a ***fetus*** in the uterus. A baby grows in the female's uterus about nine months before it is ready for birth.

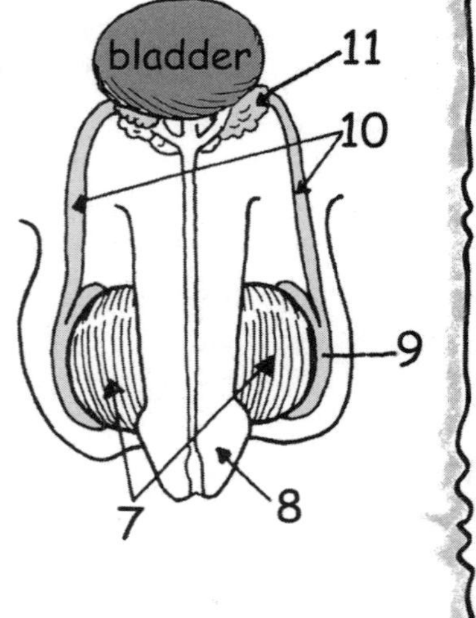

I'm Leon.
I'm Noel.
We're fraternal twins!

TWINS—How does it happen?

What causes two babies instead of one?

It can happen two different ways:

Sometimes, after a sperm has fertilized one egg, the egg splits into two at a very early stage of growth. Each part develops into a separate baby, creating ***identical twins.*** These babies are the same sex and have identical chromosomes.

Sometimes, two eggs are released at the same time, and a separate sperm fertilizes each one. These grow into two separate babies, called ***fraternal twins***. They have different chromosomes, and may be different sexes.

Genetics & Heredity

Dazed and Confused

MATERNITY WARD
Genetics? Heredity? Genes? Chromosomes? DNA? I'm really confused about all of it.
Me, too!

Heredity is the passing of traits from parents to their offspring. A ***trait*** is a characteristic of an organism (such as height or eye color).

Genes are the coded instructions in the DNA. They are the basic units of inheritance. There are hundreds of genes on each chromosome.

Genetics is the study of genes and heredity.

Chromosomes are threadlike structures in the nucleus of every cell that carry the codes for a cell's activity. There are 23 chromosomes in each sperm and each egg. When the sperm and egg combine, they give 46 chromosomes to the baby. The chromosomes are made of proteins and a chemical, *DNA* (deoxyribonucleic acid). Each DNA molecule contains codes that have instructions to control the way the cells work.

Get Sharp Tip #20

Every person has 23 sets of chromosomes in every cell.

Which sex?

Every person has two sex chromosomes. Males have an X and a Y chromosome. Females have two X chromosomes.

All eggs have an X. Half of all sperm have an X and half have a Y.

If a sperm with an X chromosome fertilizes an egg, the baby will be a girl. If a sperm with a Y chromosome fertilizes the egg, the baby will be a boy.

What color hair & eyes?

Cells carry two or more genes for each characteristic. Some genes are ***dominant***. They tend to overpower the weaker, or ***recessive*** genes.

The gene for dark hair is dominant over the gene for light hair. Brown eyes are dominant over blue eyes.

A child born to a dark-haired, brown-eyed parent and a blonde, blue-eyed parent is most likely to have dark hair and brown eyes.

Diseases & Disorders

There are many things that can go wrong in the body. Henry Hypochondriac thinks he has them all! Read about these diseases and disorders.

With this ***arthritis,*** my joints are swollen all the time.

I fell off a camel and have ***fractures***—three broken bones in my leg.

I have ***hemophilia***; my blood doesn't clot properly. It is an inherited condition.

My appendix was so inflamed, it was about to burst. The doctor said I had ***appendicitis.***

I have ***pneumonia***, an infection in my bronchioles (the smallest tubes in my lungs).

Hepatitis is my problem. It is a serious infection in my liver.

I have ***gastroenteritis***, an infection of my stomach and intestines.

I have ***influenza***, a nasty ailment caused by a virus.

I need to get treatment for ***tetanus***. It is caused by bacteria in an infected wound. The toxins in the bacteria can cause nerve paralysis.

My body is unusually sensitive to pollen. This ***allergy*** is causing my eyes to itch all the time.

I have a highly infectious disease, ***rabies.*** I got it when a raccoon bit me.

I've lost my voice. My voice box is inflamed. I have ***laryngitis***.

I can't breathe well! Due to my ***asthma***, my bronchioles are blocked, and I'm struggling for air.

This ***pyorrhea*** is an infection of my gums.

I've got ***food poisoning.*** There was bacteria in something I ate.

I have the ***mumps.*** My parotid glands are infected.

This ***sprain*** was caused when I twisted my ankle joint too far.

Defenses Against Disease

The body has a wonderful system of natural defenses to help it fight against diseases and other problems. Body defenses also help it heal from many ailments.

White blood cells can surround and digest germs.

Antibodies, made by some blood cells, attack and kill germs.

Platelets in blood help the blood clot to stop bleeding.

Acid in the stomach kills germs.

Saliva kills germs and bacteria in the mouth.

The liver and kidneys filter toxic substances out of the blood.

Clean, unbroken skin keeps germs out of the body.

Bone cells are able to multiply and fill in breaks in a bone.

A fever shows the growth of germs. It warns of illness.

Hairs in the nose filter germs out of the air before they get into the lungs.

Mucus in the nose and throat kills germs before they get farther into the body.

Eye-blinking is a reflex that keeps harmful things out of the eyes.

Coughing is a reflex that catches and expels harmful substances before they get into the lungs.

Vomiting is a reflex that removes harmful substances from the stomach.

Sneezing is a reflex that keeps dust and germs out of the respiratory tract.

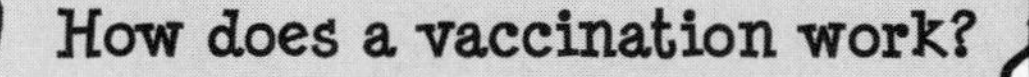

How does a vaccination work?

Dead or weakened germs are put into the body. The presence of these germs causes the body to produce antibodies that stay in the body for a long time and protect it from an attack by the live germs. Vaccinations are given by injection or by mouth.

The body's defense system can get some outside help from:

* antibiotics to slow or stop bacterial growth
* careful food inspection and preparation
* clean water programs
* good hygiene such as handwashing, dental care, and general cleanliness
* vaccines to build up immunities to disease
* surgery to repair damaged organs
* good nutrition to provide the body with substances for growth and repair
* sunshine to provide vitamin D needed by the body for proper use of calcium (to keep strong bones)

Health & Fitness

Being healthy means having a body that is working well. Being fit means you are able to use your body to do the things you want to do. There are many aspects to health and fitness. Keeping healthy and fit takes understanding and commitment. The benefits are well worth your time and energy!

Healthy Eating

Your health is definitely affected by what you eat. Make sure you are getting the nutrients you need from these food groups. Don't let fatty, sugary foods or drinks crowd the good things out of your diet. Every day, get a proper supply of . . .

. . . **protein** in lean meat, fish, seafood, eggs, dairy products, lentils, beans, and nuts. Protein supplies energy. It also helps your body build new cells and repair damaged cells.

. . . **fats** in small amounts in oils, meat, and dairy products such as low-fat milk, yogurt, cheeses, and butter. Choose vegetable fats rather than the saturated fats found in animal fats.

. . . **carbohydrates** in grains, breads, rice, cereal, pasta, fruits, and vegetables. Carbohydrates are an important source of energy. Get most of yours from healthy fruits, vegetables, and whole grains.

. . . **vitamins and minerals** from a variety of fruits, vegetables, and fresh foods. You need these to keep the systems of your body functioning properly.

. . . **fiber** to keep your digestive system healthy. You can get fiber in fruits and vegetables, whole grains, nuts, seeds, and beans.

. . . **water**. You need about two liters (eight glasses) of water every day. If you exercise or work hard and lose water by sweating, drink more!

Some Aspects of Health and Fitness

Healthy, balanced diet
Stamina
Strength
Flexibility
Cleanliness
Regular eye, ear, and dental care
Disease prevention
Avoiding dangerous behaviors
Good posture
Getting plenty of rest
Stress management

Doctor! Doctor!
Doctor, doctor! I think I swallowed a roll of film! What should I do?
Take two aspirin and we'll see what develops.

Other Good Advice

- Go easy on sugar, pastries and pastas with white flour, salt, sweet drinks, and caffeine.
- Avoid fried foods as much as possible.
- Eat food as fresh as possible.
- Maintain healthy weight by moderate, balanced, healthy eating—NOT by crash dieting!

Exercise

Your body needs three different kinds of exercise on a regular basis.

> **Get Sharp Tip #21**
> All forms of exercise help to relieve stress.

Aerobic Exercise

What is it? Aerobic exercise causes the heart to work harder or beat faster for a period of time (20 minutes or longer).

What's the benefit? It builds heart strength and stamina, and improves lung function. Aerobic exercise is also a great way to relieve stress.

What kinds of activities are aerobic? You can get aerobic benefit from any activity that keeps the body (especially large muscles) moving for a period of time: walking, jogging, running, swimming, jumping rope, cross-country skiing, biking, stair-stepping, etc.

Flexibility Exercise

What is it? Flexibility exercise is movement that keeps your body supple (able to stretch, bend, and twist in different directions).

What's the benefit? It allows the body to perform a wide range of movements without getting injured or sore. It also encourages good posture, improves balance, and reduces stress.

What kinds of activities build flexibility? Any exercises that gently bend and stretch will help increase flexibility. Yoga and other stretching movements are good.

Strengthening Exercise

What is it? Strength is the amount of force produced by muscles. Exercise that strengthens the body builds the amount of force a muscle group has.

What's the benefit? Strengthening exercise builds the size of the muscles and makes them firmer. This gives you more power in those muscles. Strong, firm muscles give you better posture and hold your organs in place well. With more muscle mass, more calories are burned in your body. This helps with weight control.

What kinds of activities build strength? To develop muscles, they need to work hard for a short period of time. Weightlifting, rowing, canoeing, or any other work or exercise that repeatedly uses upper body or lower body muscle strength will build the muscles.

Doctor, Doctor!
WAITING ROOM
What does your family think of your new hearing aids, Mr. Jones?
Oh, I haven't told them yet, Doctor. I'm having too much fun listening in on their conversations.

First Aid

First aid is immediate and temporary help given to someone who suddenly becomes ill or injured. Knowing first aid can lead you to be more conscious about your own safety. It can also make a critical, perhaps life-saving difference to you or someone else in an emergency.

I'm feeling much better now.

Ailment	Symptoms	First Aid
bleeding	blood coming from broken skin	Put direct pressure over the wound with a clean cloth. Get medical help for severe bleeding.
wound	broken skin	Wash with soap and water; cover with sterile bandage.
head wound	dizziness, vomiting, bleeding, unconsciousness, enlarged pupils	Lay victim down with pillow under head. Call for medical help immediately.
shock	pale, cold, clammy skin, irregular pulse, shallow breathing	Keep victim lying down with feet slightly raised. Cover with blankets. Give salt water every 15 minutes if conscious. Get medical help.
fracture	severe pain, swelling	Call for medical help. Apply ice to area. Do not move the victim or injured area.
burns	reddening skin, blistering	For burns that redden the skin, place under cold water. Get medical attention for other burns.
poisoning	stomach sickness, awareness that poison has been drunk	Call a poison control center. Identify the substance. Find the container. It might tell an antidote for the substance.
animal bite	broken skin, possible bleeding	Wash with soap and water. Apply ice to reduce swelling. Check the animal for rabies.
insect sting	itching, swelling	Wash with soap and water. Apply ice to reduce swelling. Scrape the stinger out of the skin gently with a tweezers, knife, or fingernail.
snakebite	pain, swelling, purple color, weakness, nausea, rapid pulse, shortness of breath	Keep victim calm with no movement. Keep bitten area below heart level. Get victim to hospital immediately.
frostbite	tingling or numbness in nose, feet, hands, or ears; pain; itching; redness	Wrap affected part in a warm blanket or soak in warm (not hot) water. Drink hot fluids.
heat exhaustion	dizziness, headache, skin looks pale and moist	Drink liquids to prevent heat exhaustion. Move victim to a cool spot. Give plenty of water and fruit juice. Give salt water every 15 minutes.
hypothermia	tiredness, shivering, chill, mental confusion, loss of muscle coordination, slurred speech, rigid muscles, loss of consciousness	Treat immediately. Get victim out of wet clothing; cover with blankets; get to a warm place. If victim is conscious, give a warm drink. Get victim to a doctor.
blockage of airway	can't talk, choking, bluish skin, dilated pupils	Try the Heimlich maneuver (abdominal thrust). Call for help immediately.
fainting	pale, moist skin; falling; loss of consciousness	Lay victim flat on the ground and raise the legs. Keep the victim resting for 10 minutes or more.
object in eye	Soreness, swelling, tears	Do not rub the eye. Pull the upper eyelid over the lower eyelid. Wash the eye with an eyewash kit.
nosebleed	sudden bleeding from nose	Keep victim quiet, sitting and leaning forward. Pinch the nostrils together. Apply something cold to nose and face.

GET SHARP

in

PHYSICAL SCIENCE

Matter

Matter is the material that makes up all things in the universe. All matter is made up of smaller particles.

An ***atom*** is the smallest piece of material that can exist on its own.
Every kind of matter is made up of one or more kinds of atoms.

Molecules are combinations of atoms.

Subatomic particles are even smaller than atoms.

A ***nucleus*** is at the center of each atom. It is made of ***protons*** and ***neutrons***.

Electrons orbit the nucleus.

The different orbits in an atom can hold different numbers of electrons. The orbit closest to the nucleus holds two. The second orbit holds eight. The third can hold eighteen. However, any orbit that is the last orbit in an atom will hold only eight electrons.

Protons carry a positive electric charge. Electrons carry a negative charge. Neutrons carry no charge.

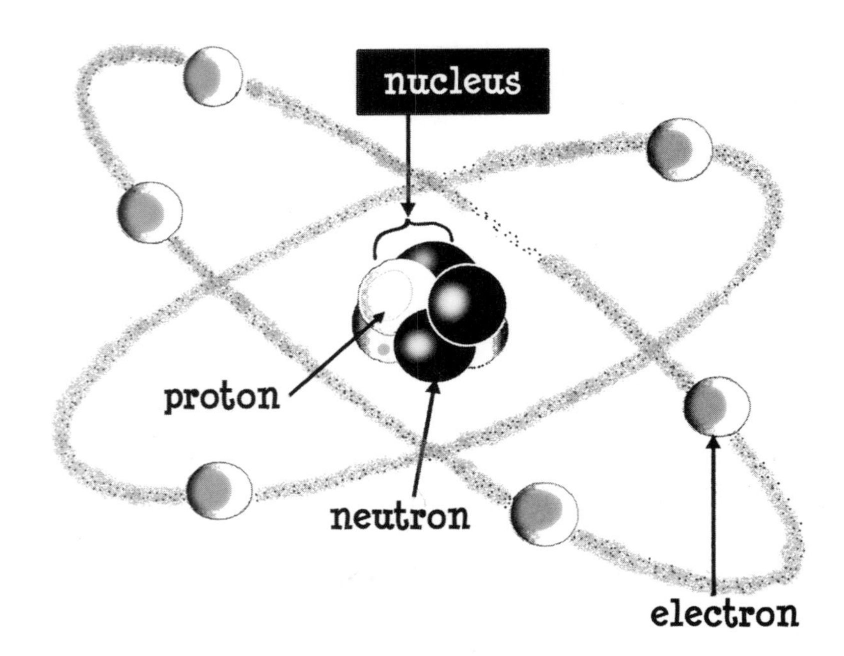

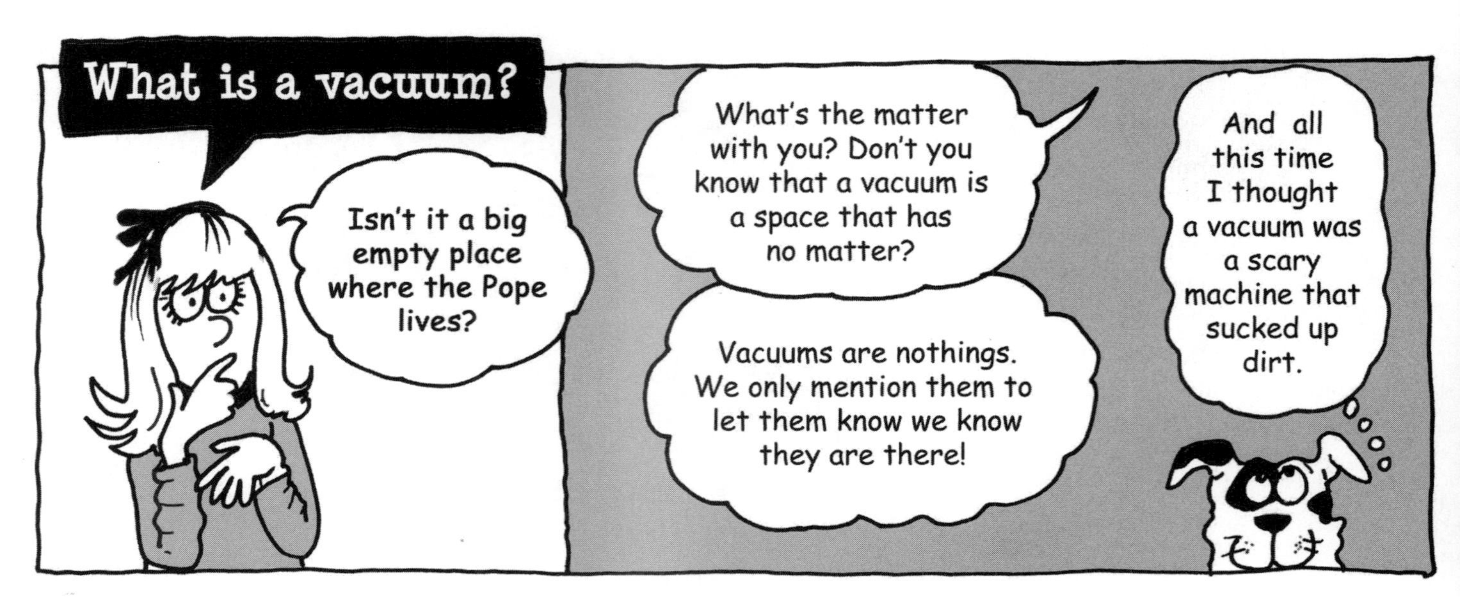

States of Matter

There are three states in which matter exists under normal conditions:

Solids
have a definite size (volume) and shape. The particles are packed together tightly and arranged in a regular pattern.

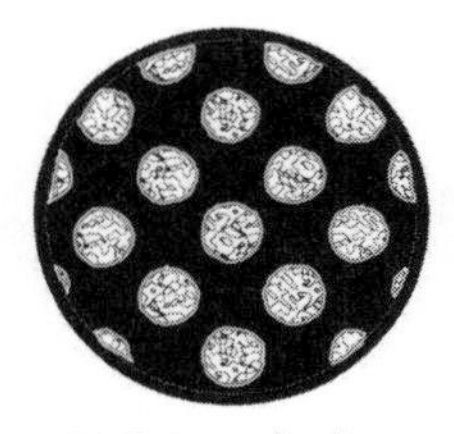

Liquids
have a definite size (volume) but no definite shape. The particles are more active and farther apart than in a solid.

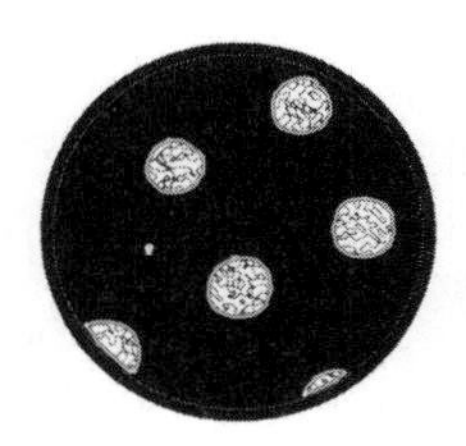

Gases
have no definite size or shape. A gas fills whatever container it occupies. The particles move freely and are far from each other.

Changes in States

Freezing is a change from a liquid to a solid state, caused by a lowered temperature. Different substances have different freezing points. Water freezes at 32° F or 0° C.

Melting is a change from a solid to a liquid state, caused by a raised temperature.

Evaporation is a change from a liquid to a gaseous state, caused when the liquid is heated to its boiling point. Different substances have different boiling points. Water boils (and evaporates) at 212° F or 100° C.

Condensation is a change from a gaseous to a liquid state, caused by a lowering of temperature.

Sublimation is a change from a solid to a gas without going through a liquid state.

Properties of Matter

Substances have different properties. ***Physical properties*** are characteristics that can be observed without changing the chemical makeup of the substances. ***Chemical properties*** are characteristics that describe how a substance will react with another substance.

Physical Properties

Mass is the amount of matter in an object. It is measured in units used to find weight (such as pounds or grams). All matter has mass and takes up space.

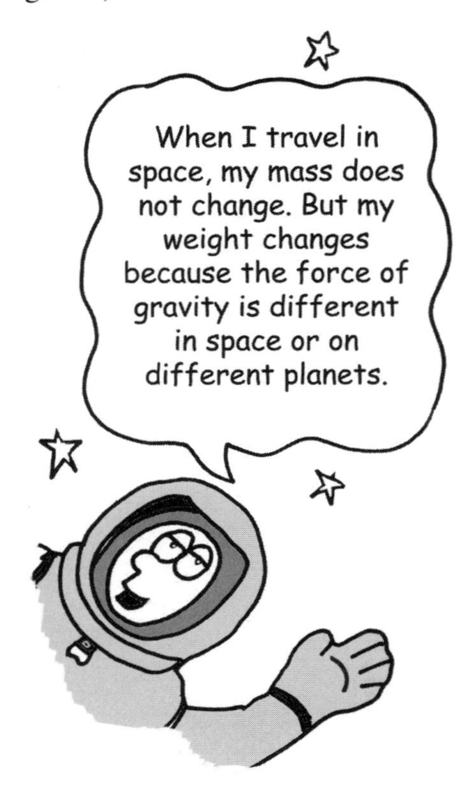

When I travel in space, my mass does not change. But my weight changes because the force of gravity is different in space or on different planets.

Weight is the force of gravity pulling on an object. Gravity is expressed in Gs, with gravity on Earth having a value of 1G. The weight of an object is its mass multiplied by the force of gravity (mass x 1G on Earth). On Earth, an object with a mass of 40 pounds will weigh 40 pounds.

Density is the amount of matter packed into a given unit of mass. Density is the mass of the material divided by its volume. It is expressed or measured in cubic units.

Viscosity is the property of a liquid that describes how it pours. Syrup has a greater viscosity than water.

Freezing point and boiling point are the temperature at which the liquid form of the substance will become solid or the liquid form will become a gas.

Ability to conduct heat (or resist heat) is another property. Substances that conduct heat are called ***conductors***. ***Insulators*** are substances that slow the movement of heat.

Magnetism is the property of attracting certain other substances. Most magnetic substances are metal.

Color, shape, size, odor, and hardness are some other physical properties.

Chemical Properties

A chemical property cannot be observed from looking at the substance. It shows up when the substance interacts with something else. Here are a few examples:

- Charcoal combines with oxygen during burning to form carbon dioxide.
- Sodium will combine with chlorine to form sodium chloride, an edible salt.
- By themselves, each of those substances are not safe to eat!
- When iron combines with oxygen in the air, the metal rusts. ***Corrosion***, or the ability to rust, is a chemical property. Some substances do not rust. This is a chemical property also.

Fluids & Gases

Fluids are substances that flow. A fluid pushes on objects with a force called a buoyant force. This force pushes on ships floating on the ocean and on airplanes flying through the air. The amount of buoyant force is related to the amount of the fluid that is displaced by the object and the weight of the object.

Gases exert force on everything around them. The amount of force a gas exerts per unit of area is called pressure. Busy molecules moving fast inside the container cause the pressure. Every time a molecule bumps against the surface, it exerts a force.

These are some special scientific rules and laws that explain the behavior of fluids and gases.

Get Sharp Tip #22

An object will **sink** if the buoyant force is less than the force of gravity pulling on the object. An object will **float** if the buoyant force is greater than the force of gravity pulling on the object.

These scientific principles make a lot of sense.

Pascal's Principle

Water pushes up on a floating ship with a force equal to the weight of the ship.

Archimedes' Principle

If you squeeze a plastic bottle of water, the pressure you put on the water will be equal throughout the water. *(There won't be more pressure where you are squeezing!)*

Bernoulli's Principle

In places where air is moving slowly, the air pressure will be high. In places where air is moving fast, the pressure will be low.

Boyle's Law

If you decrease the amount of gas in a balloon, the pressure will increase. *(This is true as long as you do not change the temperature of the gas.)*

Charles' Law

If you increase the temperature of a gas in a balloon, the volume will increase. *(This is true as long as the pressure stays the same.)*

Elements

An ***element*** is a substance that contains only one kind of atom. An element cannot be broken down by physical or chemical means. There are 103 elements that have been officially named. Most of them occur naturally. Each named element has been given a symbol. There are two main groups of elements: ***metals*** and ***nonmetals.*** Some elements have characteristics of both and are called ***transition elements***.

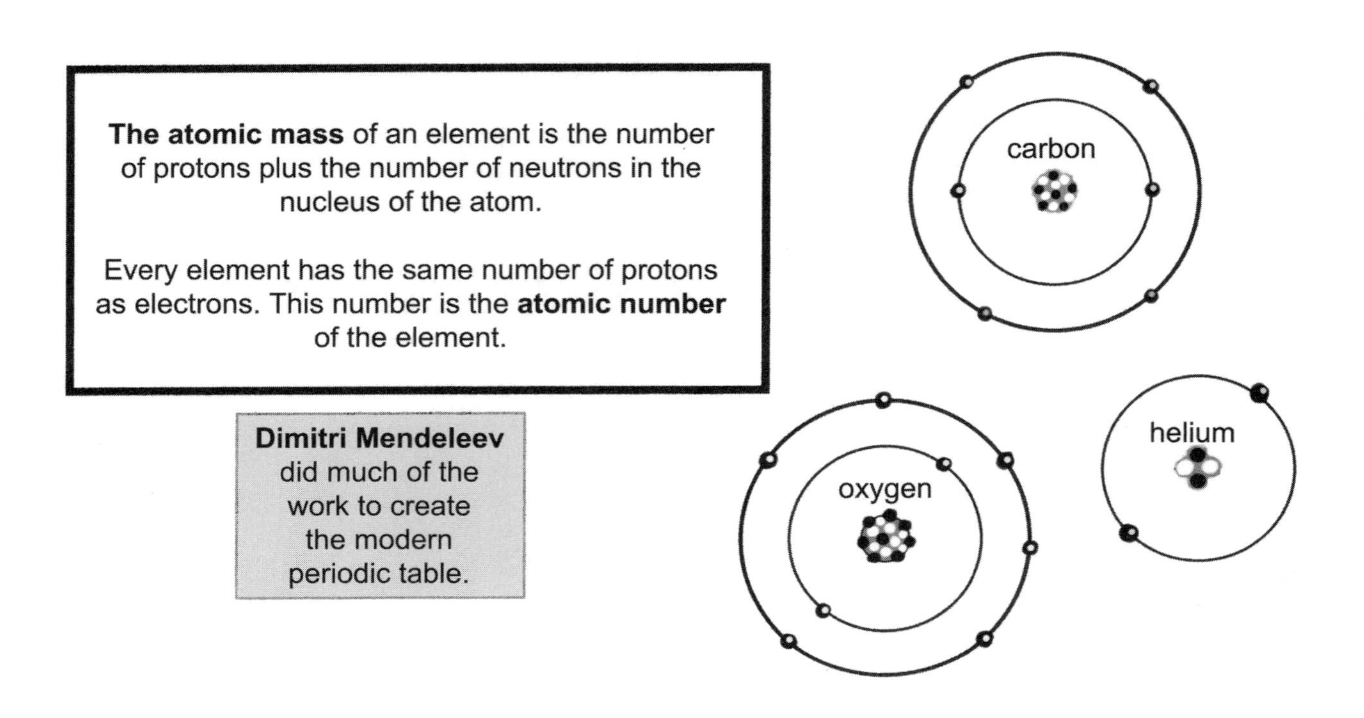

The atomic mass of an element is the number of protons plus the number of neutrons in the nucleus of the atom.
Every element has the same number of protons as electrons. This number is the **atomic number** of the element.
Dimitri Mendeleev did much of the work to create the modern periodic table.
carbon
oxygen
helium

The Periodic Table

The periodic table shows information about the elements. On the table, elements are arranged in rows, called ***periods,*** in order of their atomic numbers. They are also arranged in columns, called ***groups***. Each group contains elements with similar properties. You can use the periodic table to learn about the elements. Here are a few things the table tells you:

- *Argon is a gas.*
- *Cadmium is a heavier atom than magnesium.*
- *Gold has 79 electrons.*
- *Hydrogen has no neutrons.*
- *The symbol for iron is Fe.*
- *P is not the symbol for potassium.*
- *Aluminum has 14 neutrons.*
- *Barium has metal properties.*

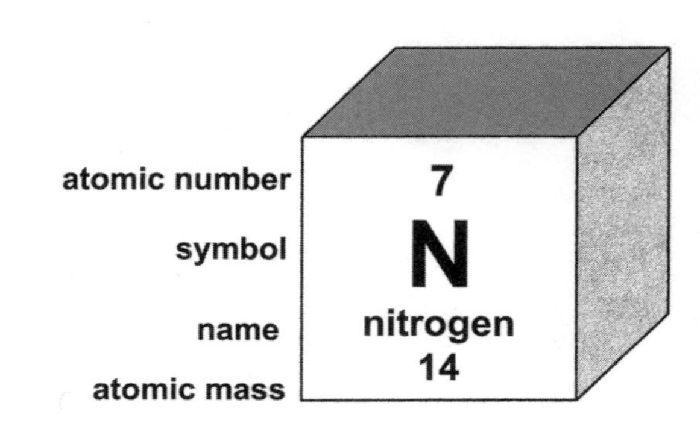

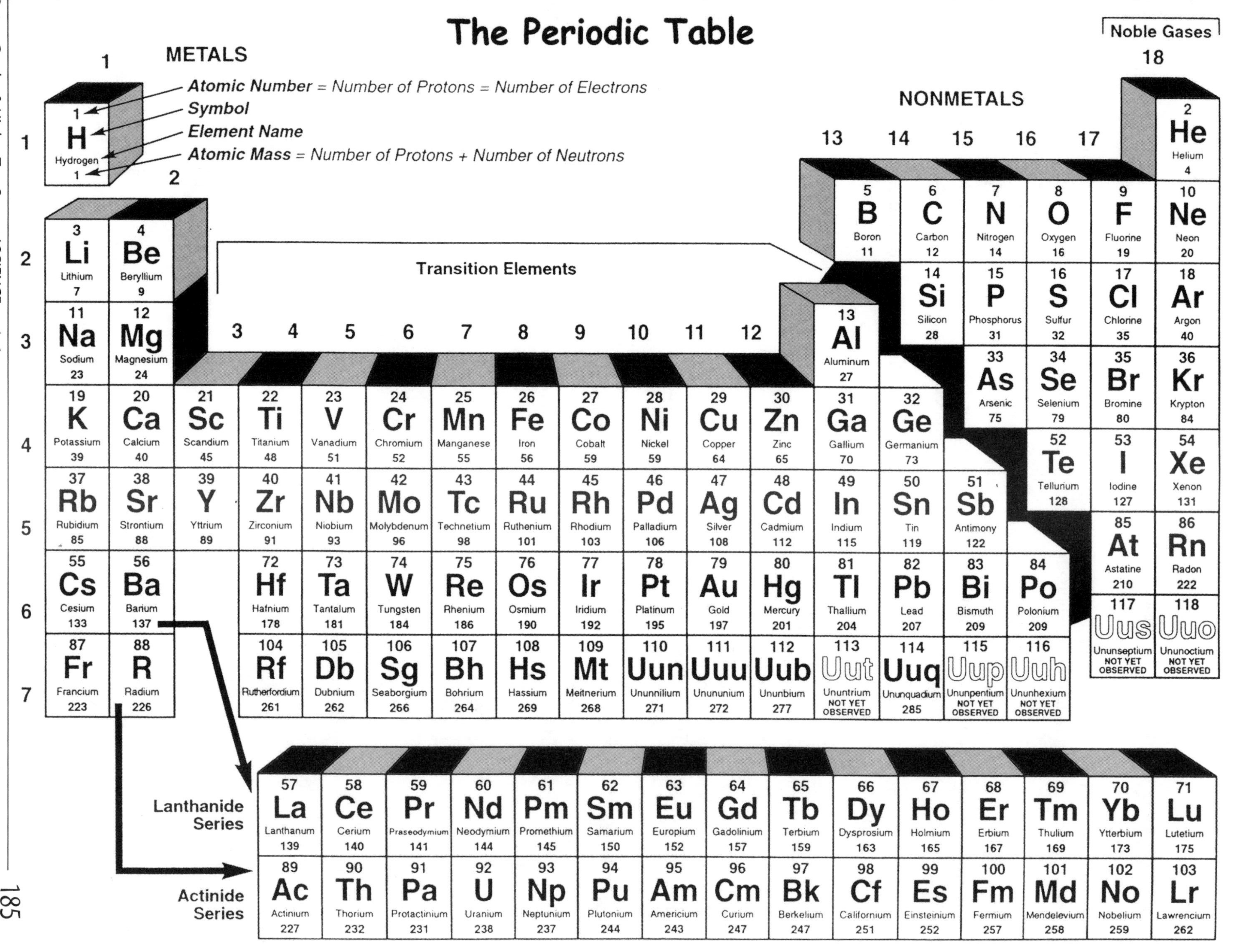

The Periodic Table

Key (shown on the Hydrogen box):
- *Atomic Number = Number of Protons = Number of Electrons*
- *Symbol*
- *Element Name*
- *Atomic Mass = Number of Protons + Number of Neutrons*

METALS | Transition Elements | NONMETALS | Noble Gases

Period	1	2	3	4	5	6	7	8	9	10	11	12	13	14	15	16	17	18
1	1 H Hydrogen 1																	2 He Helium 4
2	3 Li Lithium 7	4 Be Beryllium 9											5 B Boron 11	6 C Carbon 12	7 N Nitrogen 14	8 O Oxygen 16	9 F Fluorine 19	10 Ne Neon 20
3	11 Na Sodium 23	12 Mg Magnesium 24											13 Al Aluminum 27	14 Si Silicon 28	15 P Phosphorus 31	16 S Sulfur 32	17 Cl Chlorine 35	18 Ar Argon 40
4	19 K Potassium 39	20 Ca Calcium 40	21 Sc Scandium 45	22 Ti Titanium 48	23 V Vanadium 51	24 Cr Chromium 52	25 Mn Manganese 55	26 Fe Iron 56	27 Co Cobalt 59	28 Ni Nickel 59	29 Cu Copper 64	30 Zn Zinc 65	31 Ga Gallium 70	32 Ge Germanium 73	33 As Arsenic 75	34 Se Selenium 79	35 Br Bromine 80	36 Kr Krypton 84
5	37 Rb Rubidium 85	38 Sr Strontium 88	39 Y Yttrium 89	40 Zr Zirconium 91	41 Nb Niobium 93	42 Mo Molybdenum 96	43 Tc Technetium 98	44 Ru Ruthenium 101	45 Rh Rhodium 103	46 Pd Palladium 106	47 Ag Silver 108	48 Cd Cadmium 112	49 In Indium 115	50 Sn Tin 119	51 Sb Antimony 122	52 Te Tellurium 128	53 I Iodine 127	54 Xe Xenon 131
6	55 Cs Cesium 133	56 Ba Barium 137		72 Hf Hafnium 178	73 Ta Tantalum 181	74 W Tungsten 184	75 Re Rhenium 186	76 Os Osmium 190	77 Ir Iridium 192	78 Pt Platinum 195	79 Au Gold 197	80 Hg Mercury 201	81 Tl Thallium 204	82 Pb Lead 207	83 Bi Bismuth 209	84 Po Polonium 209	85 At Astatine 210	86 Rn Radon 222
7	87 Fr Francium 223	88 R Radium 226		104 Rf Rutherfordium 261	105 Db Dubnium 262	106 Sg Seaborgium 266	107 Bh Bohrium 264	108 Hs Hassium 269	109 Mt Meitnerium 268	110 Uun Ununnilium 271	111 Uuu Unununium 272	112 Uub Ununbium 277	113 Uut Ununtrium NOT YET OBSERVED	114 Uuq Ununquadium 285	115 Uup Ununpentium NOT YET OBSERVED	116 Uuh Ununhexium NOT YET OBSERVED	117 Uus Ununseptium NOT YET OBSERVED	118 Uuo Ununoctium NOT YET OBSERVED

Lanthanide Series	57 La Lanthanum 139	58 Ce Cerium 140	59 Pr Praseodymium 141	60 Nd Neodymium 144	61 Pm Promethium 145	62 Sm Samarium 150	63 Eu Europium 152	64 Gd Gadolinium 157	65 Tb Terbium 159	66 Dy Dysprosium 163	67 Ho Holmium 165	68 Er Erbium 167	69 Tm Thulium 169	70 Yb Ytterbium 173	71 Lu Lutetium 175
Actinide Series	89 Ac Actinium 227	90 Th Thorium 232	91 Pa Protactinium 231	92 U Uranium 238	93 Np Neptunium 237	94 Pu Plutonium 244	95 Am Americium 243	96 Cm Curium 247	97 Bk Berkelium 247	98 Cf Californium 251	99 Es Einsteinium 252	100 Fm Fermium 257	101 Md Mendelevium 258	102 No Nobelium 259	103 Lr Lawrencium 262

Compounds

Compounds are formed when two or more elements are joined together chemically. The molecules form a similar element each time the compound is formed. When compounds form, bonds between atoms are rearranged. Atoms of different kinds lose, gain, or share electrons to form bonds between them.

The elements in a compound cannot be separated by breaking, melting, freezing, filtering, evaporating, or any other physical means. When elements join in a compound, each element no longer has the same properties as it did before. The compound has different properties than the original elements.

Each compound has a chemical formula. The formula shows the number of atoms of each element in a molecule of the compound.

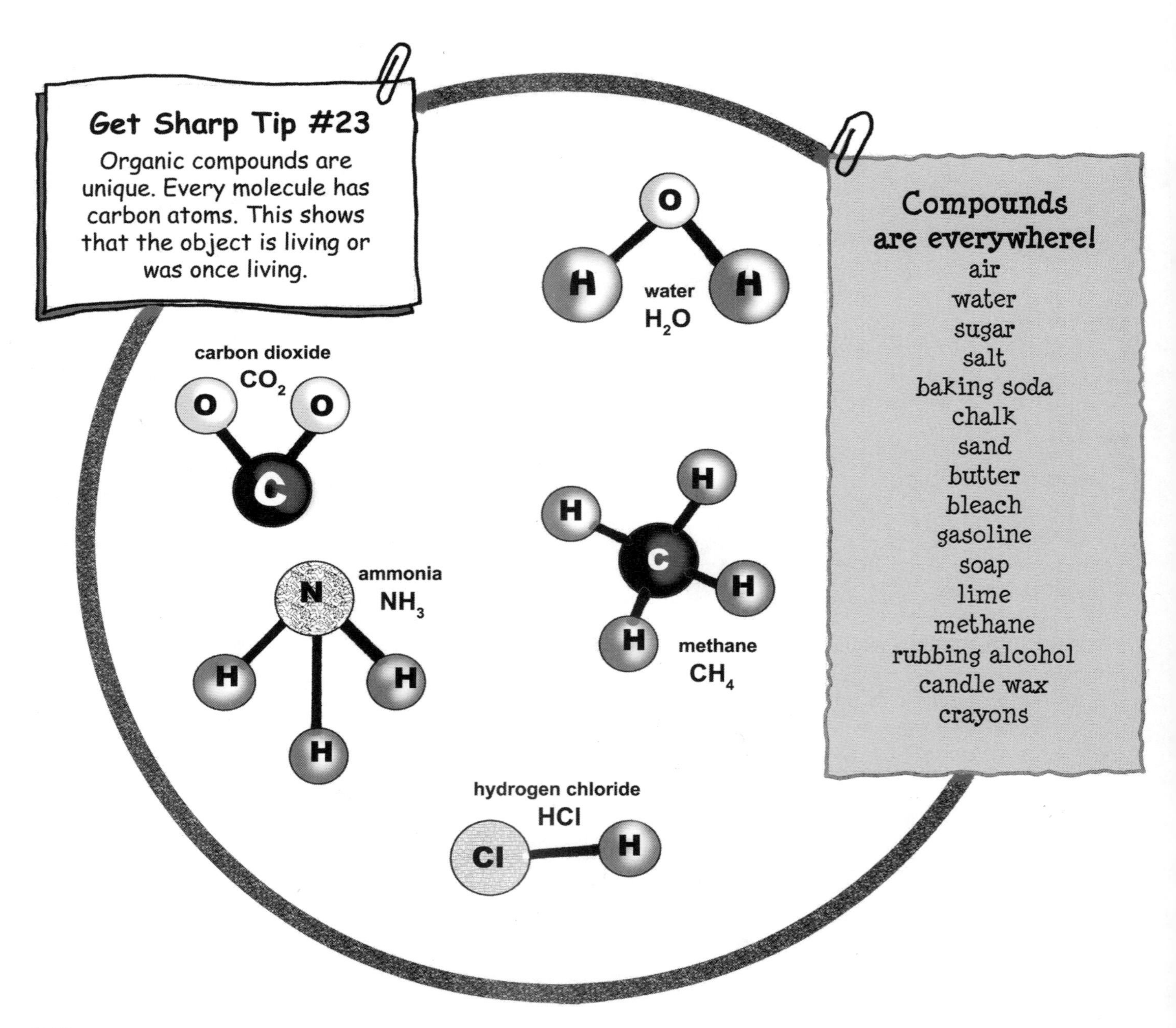

Get Sharp Tip #23
Organic compounds are unique. Every molecule has carbon atoms. This shows that the object is living or was once living.
water
H_2O
carbon dioxide
CO_2
ammonia
NH_3
methane
CH_4
hydrogen chloride
HCl
Compounds are everywhere!
air
water
sugar
salt
baking soda
chalk
sand
butter
bleach
gasoline
soap
lime
methane
rubbing alcohol
candle wax
crayons

Some Common Compounds

Compounds are a part of everyday life.
Here are a few of the compounds that get regular use.

Common Name	Formula
marble *(calcium carbonate)*	$CaCO_3$
bleach *(sodium hypochlorite)*	$NaClO$
salt *(sodium chloride)*	$NaCl$
candle wax	CH_2
baking soda *(sodium hydrogen carbonate)*	$NaHCO_3$
sugar *(sucrose)*	$C_{12}H_{22}O_{11}$
methane	CH_4
propane	C_3H_8
chalk *(calcium carbonate)*	$CaCO_3$
ammonia	NH_3
sand *(silicon dioxide)*	SiO_2
gasoline *(octane)*	C_8H_{18}
hydrogen peroxide	H_2O_2
Epsom salts	$MgSO_47H_2O$
carbon dioxide	CO_2
milk of magnesia *(magnesium hydroxide)*	$Mg(OH)_2$
Freon *(dichlorodifluoromethane)*	CF_2Cl_2
lime *(calcium oxide)*	CaO
soap *(glycerin)*	$C_3H_8O_3$
copper sulfate	$CuSO_4$
antifreeze *(ethylene glycol)*	$C_2H_6O_2$
water	H_2O

Mixtures

A ***mixture*** is a combination of two or more substances blended together without a chemical reaction. In a mixture,

. . . each substance keeps its own properties.
. . . there is no chemical reaction.
. . . there is no repeated chemical makeup or chemical formula.
. . . substances can be separated.

Two Kinds of Mixtures

A **heterogeneous mixture** is a mixture in which the particles are not spread evenly throughout.

Milkshakes, dirt, and salad dressings with oil, vinegar, and spices are heterogeneous mixtures.

tea, milkshake

A **suspension** is a heterogeneous mixture with particles large enough to be seen by the eye or with a microscope. A suspension looks cloudy. The particles in a suspension are not dissolved; they will settle out from the force of gravity. Shaking or stirring will suspend the particles again. The particles can be separated out with filter paper.

Orange juice with pulp is a suspension.

orange juice with pulp

cheese

A **colloid** is a heterogeneous mixture with particles of a size between that of a solution and a suspension. The particles do not settle out with gravity, and cannot be filtered out.

Fog, clouds, cheese, jam, and whipped cream are colloids.

A **homogeneous mixture** is a mixture in which particles of one substance are spread evenly throughout the other substance.

Tea, root beer, vinegar, and lemon juice (without pulp) are homogeneous mixtures.

A **solution** is a homogeneous mixture with very tiny particles of a substance spread evenly throughout. The particles will not settle out when the mixture sits for a while, and a filter cannot separate the particles.

Salt water is a solution.

salt water

Separating Mixtures

Substances combined in a mixture can be easily separated by methods such as:

. . . separating metal out of a liquid by using a magnet

. . . sorting by hand

. . . shaking smaller particles through a sieve

. . . settling out solid particles by letting the mixture stand

. . . passing the substance through filter paper

. . . distilling to separate two liquids with different boiling points

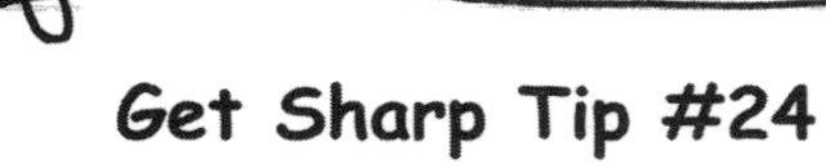

Get Sharp Tip #24

In distillation, one liquid in a mixture evaporates sooner than the other. The vapor of that liquid is collected and cooled back into a liquid.

You Can Make Gobs of Glorious Goop
(and amaze your friends and relatives)

Is it a solid?
Is it a liquid?
This suspension has characteristics of both. Sometimes it acts like a solid. Sometimes it acts like a liquid.

Put $1\frac{1}{2}$ C of cornstarch into a bowl. Carefully stir in 1 C of water. It will be hard to stir. Add 2 drops of green food coloring.

Put some of the goop into your hands. Roll a ball and squeeze it until it is hard.

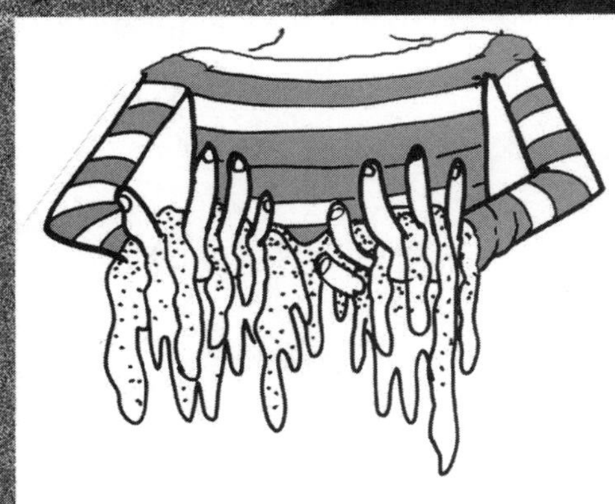

Now, stop squeezing and let it run through your fingers!

Solutions

A ***solution*** is a homogeneous mixture in which one substance is dissolved in another. Two or more substances are mixed uniformly. The ***solute*** is the substance being dissolved. The ***solvent*** is the substance in which the solute is dissolved.

An **aqueous solution** is a solution in which water is the solvent. One or more solids, liquids, or gases are dissolved in the water. *(Examples: soda pop and cranberry juice)*

A **gaseous solution** is a solution in which a solid, liquid, or gas is dissolved in a gas. *(Example: air)*

A **solid solution** is a solution in which a solid, liquid, or gas is dissolved in a solid. *(Example: steel, which is carbon dissolved in iron)*

Chef D'Jour's Cooking School

Today's Lesson:
How To Make Cactus Juice
125 mL / 500 mL =
.25% x 100% = 25%.
Only 25% of this is pure cactus juice!

The **concentration** is the amount of a solute that is dissolved per unit of solvent. This cactus drink has 500 mL total in the container. 375 mL of the mixture is water. The rest is cactus juice.

To find the concentration, divide the amount of the solute by the total amount of the mixure. Then multiply that by 100%.

Any of these will cause the solute to dissolve faster:

1. Heat the mixture.
2. Crush the solid into smaller pieces.
3. Shake or stir the mixture.

Solubility

Solubility is the measure of the amount of solute that can be dissolved in a specific amount of solvent at a specific temperature.

Get Sharp Tip #25
Solubility can be increased by raising the temperature.

An **unsaturated solution** is a solution that can hold more solute.

A **saturated solution** is a solution that holds all the solute it can.

A **supersaturated solution** holds more solute than normal at this temperature. One more drop of solute will crystallize immediately.

Acids, Bases, & Salts

Most compounds are one of these: an acid, a base, or a salt.

Antacids use a basic solution to neutralize stomach acid that causes indigestion.

An **acid . . .**

. . . has a strong smell and a sour taste
. . . reacts with metals
. . . contains hydrogen
. . . can burn or sting skin

Acidic Substances
vinegar
stomach acid
lemons
oranges
grapefruit
buttermilk
sulfuric acid
bee stings
car batteries

A **base . . .**

. . . tastes bitter
. . . feels slippery
. . . can burn the skin
. . . can dissolve things
. . . can be poisonous

Bases that are soluble in water are called ***alkali***. When you find the alkalinity of a substance, you are finding out how strong the base characteristic is.

Alkaline
(base)
Substances
deodorant
antacids
soap
ammonia
drain cleaner
oven cleaner

Salts are products that result when an acid and base ***neutralize*** each other. When the right quantities of an acid and a base are combined, a neutral salt results. Neutralization always produces water and a salt.

A solution or paste of baking soda (base) can relieve the pain of a bee sting (acid).

Testing for Acids and Bases

You can test a solution to find out whether it is an acidic or alkaline (basic) by using an ***indicator***, such as litmus paper or phenolphthalein. A good indicator will also tell the level of acid or base. Indicators are generally organic compounds that turn different colors in the presence of acids or bases. The indicator's shades of color will vary depending on the strength of the acid or base.

Scientists use a range of numbers, called a ***pH scale***, to tell how acidic or basic a solution is. On the scale, acids range from 1–6 and bases range from 8–14. A solution that tests 7 is neutral.

1 2 3 4 5 6 7 8 9 10 11 12 13 14

strongly acidic — weakly acidic — neutral — weakly alkaline — strongly alkaline

Indicators

Litmus paper

An ***acidic solution*** turns ***blue*** litmus paper ***red***.

A ***basic solution*** turns ***red*** litmus paper ***blue***.

A ***neutral solution*** will not change the color of either paper.

Phenolphthalein

An ***acidic solution*** turns it colorless.

A ***basic solution*** turns it bright pink.

Some pH Values

Substance	pH
milk of magnesia	10.5
stomach juices	1.5
ammonia	11
orange juice	3.5
apple juice	3
milk	6.7
water	7
blood	7.5
sea water	8.5
vinegar	2.8

Red Cabbage

An indicator can be made from the juice of a red cabbage. Make your own.

How to Make Your Own Red Cabbage Indicator

1. Grate $\frac{1}{2}$ a red cabbage into a bowl.
2. Add just enough water to cover the cabbage.
3. Let the cabbage stand in the water. Stir it occasionally.
4. When the water is deep red, use a large slotted spoon to remove as much cabbage as you can.
5. Pour the remaining juice through a strainer to get the rest of the cabbage pieces out.
6. Store the juice in a jar with a lid. This juice is your indicator.
7. Put small samples of substances in little dishes or jars.
8. Spoon a few drops of the indicator on each sample.
   If the sample turns **pink,** the substance is an **acid.**
   If the sample turns **green** or **blue,** the substance is a **base.**

Some Things to Test

lemon juice
baking soda
baking powder
vinegar
orange juice
buttermilk
antacid liquid
antacid tablet
laundry soap
soda pop
egg white
cranberry drink
tomato
cottage cheese
tea
coffee
cooking water
milk of magnesia

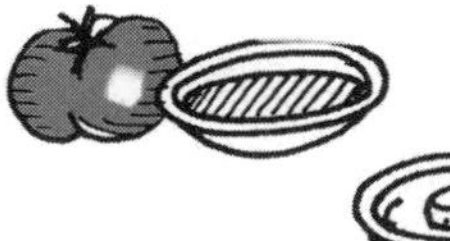

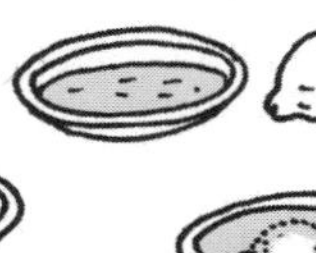

Changes in Matter

There are many ways that matter changes. All changes fall into two categories.

Physical Changes

A ***physical change*** is a process that does not change the chemical makeup of a substance. After the change, the material still has the same properties. Only the size, shape, color, state, or location has changed.

blowing a dandelion
breaking glass
separating rocks from sand
evaporating water
melting snowman
freezing popsicles
rain condensing from clouds
chocolate candy melting in the sun
clothes drying in a dryer
blowing a glass sculpture
mixing up a milkshake
slicing bread
making chocolate milk
whipping cream
squeezing oranges for juice
clothes drying in the sun
milk evaporating

The shape of the dandelion is changed. The water in the wet clothes will change from liquid to gas *(evaporation)*. The warm sun is changing the state of the chocolate (melting) and of the milk *(evaporation)*.

Chemical Changes

A ***chemical change*** does involve a change in the chemical makeup of the substance. After the change, it no longer has the same properties. It has become a new substance.

baking a cake
burning a candle
striking a match
frying an egg
milk turning sour
wood burning in campfire
shooting off fireworks
bleaching hair
garbage rotting
a bicycle rusting
an old log decomposing
plants making oxygen
toast getting crisp and brown
food digesting
bread molding
photosynthesis
fading color in fabrics

This rotting wood is dangerous!
crack
FACT!
In any change of matter (physical or chemical), some kind of energy must be applied to the substance in order for the change to occur.

Motion & Force

If you want to ride a skateboard, relocate a tent, lift weights, pull up your socks, open a jar, fly a kite, or wiggle your ears, you need a force. Nothing moves until a force acts on it. Nothing stops moving until a force acts on it.

Motion is a change in position of some object or substance.

A **force** is a push or pull acting on something. A force is needed to start a motion, stop a motion, make something go faster or slower, or change directions of object or body. Some forces act directly to push or pull an object. Other forces, such as gravity and magnetism, act at a distance.

Get Sharp Tip #26
The unit used for measuring force is the newton (N). A force of 9.8 N pulls upon each kilogram of mass in an object at rest on Earth.

Force makes it possible to :

push lower mix throw
bend pull lift squeeze stretch
raise twist press tug launch wiggle
shake jerk toss hurl haul

Gravity is the basic force in the universe. Every body in the solar system has a force that pulls things to itself. That's gravity! On Earth, it keeps people and objects from flying off into space. An object's weight depends on the force of gravity. The pull of gravity is different on different bodies in space, so weight varies on different planets or moons.

Remember: Don't let gravity get you down!

Centrifugal force and centripetal force are two forces acting on the object that spins around a second object. Centrifugal force pushes the object outward. Centripetal force pulls the object inward.

Get Sharp Tip #27

The more mass an object has, the greater will be the force of gravity pulling on it.

Net Force

When the forces acting on an object are unbalanced, a ***net force*** results. One force is stronger than the other. A net force is needed to change the motion of an object (to change its speed or direction). During a tug of war, there must be a net force for one team to pull some of the other team members across the line.

Balanced Forces

When something is not moving, the forces on it are the same. Two forces are acting on the object equally from opposite directions. If something isn't moving, the forces are always ***balanced***. During a tug of war, the forces are balanced when neither team is able to pull any of the other team members across the line.

Science Fact

Pressure is a force that presses evenly on a surface. Air exerts pressure all over Earth's surface. Its force is equal to 1 kilogram per square centimeter.

May the Force Be With You!

Zelda lies down on a puffy mattress.

The force of gravity on her mass pulls her down. The mattress bends from the force. (A net force is operating.)

The force of the mattress pushes up against her. She stops sinking into the mattress when the force of the mattress pushing equals the force of gravity pulling down on her. (The forces are balanced.)

More Things to Know About Motion & Force

Friction is the force that opposes motion between two touching surfaces.

A canoer stops paddling. The canoe keeps moving ahead for a while, but eventually friction causes it to stop.

Speed is the rate of motion of a body. It is expressed in distance per units of time.

A fast train might travel at a speed of 100 miles per hour.

Velocity is speed in a particular direction.

A horse gallops around a circle at 30 miles per hour. His speed is pretty fast, but his velocity is zero because he ends up where he started. The horse does not make progress in any direction.

Some Facts About Carly's Canoe:

Inertia keeps the canoe sitting still until someone paddles it. The harder it's paddled, the greater the velocity (speed) will be.

The boat accelerates as it gets moving. It picks up more momentum with each second.

The sleek shape reduces the air resistance against the canoe.

Acceleration is the rate of change in velocity when the velocity increases.

A car accelerates from 20 mph to 40 mph.

Deceleration is the rate of change in velocity when the velocity decreases.

It decelerates from 40 mph to 0 mph.

Inertia is the property of a body to resist any change in velocity.

A child pulls a sled along behind her on the snow. When she stops, the sled keeps moving and hits the back of her legs.

Momentum is the quantity of motion in a moving body. It is found by multiplying the mass and velocity.

A soccer ball kicked by an adult will have greater momentum than a ball rolled by a child; even though the balls have the same mass, the velocity is greater.

Resistance is any opposition that slows something down or prevents movement. It is a form of friction.

When you ride your bike into a strong wind, the wind resistance slows your velocity.

Laws of Motion

In 1687, an English scientist named Sir Isaac Newton published three rules that describe the ways force changes motion. These are known as ***Newton's Laws of Motion***.

Newton's Laws of Motion

Law #1

Every object at rest remains at rest, and every object in motion continues moving in a straight line at a steady rate, unless a force acts on it. This is the principle of inertia.

Force is needed to overcome inertia! The canoe won't move until it gets some kind of push or pull.

Law #2

The amount of force needed to change the speed of an object depends on the mass of the object and the amount of acceleration (or deceleration) needed. This is the principle of accelerated motion.

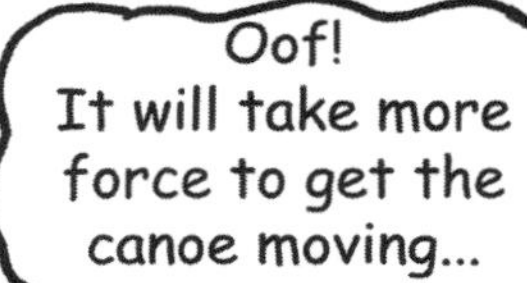

Law #3

To every action there is an opposite and equal reaction. When an object is pushed or pulled by a force in one direction, another force pushes or pulls on it with equal strength from the opposite direction.

Forces come in pairs, called action-reaction pairs.

Energy

Energy is everywhere. We can feel it as heat, see it as light, and hear it as sound. We use energy that is caused by movement of wind and water. We use energy produced by electricity or by chemical reactions inside and outside our bodies. ***Energy*** is the ability of matter to do work or to make a change in itself or its environment. Every object has energy because of its position or motion. Energy has many forms, and it can change from one form to another.

Kinetic energy is the energy of motion. An object's kinetic energy depends on its mass and velocity (speed). Objects with greater velocity or mass have more kinetic energy.

A moving bus or jumping frog has kinetic energy.

Potential energy is stored energy. Every resting object has potential energy. This means that if the object were sent into motion, it would have kinetic energy.

A roller coaster stopped at the top of a steep hill has potential energy.

Mechanical energy is potential and kinetic energy combined in actions such as lifting, pushing, pulling, stretching, or bending.

You can see mechanical energy if you watch kids pulling taffy.

Work is the transfer of energy that results from motion.

Picking up a rock is work.

Nuclear energy is energy stored in the nucleus of every atom. This energy is extremely powerful. Most nuclear energy produced comes from the atoms of certain elements such as uranium.

This energy is released by fission (splitting of the atoms) or fusion (joining together nuclei of atoms).

Radiant energy is energy that can travel through space in the form of waves.

The Sun gives off radiant energy.

Chemical energy is energy released during a chemical reaction.

The digestion of food is a chemical process that releases energy for body activity.
Burning of wood is a chemical reaction that releases energy as heat and light.

Thermal energy is the total energy in all the particles of an object.
The amount of energy depends on the temperature of the object.

Boiling soup has more thermal energy than an ice cube.

Solar energy is energy that is trapped from the Sun.

Solar energy can be stored and turned into electricity.

The Law of Conservation of Energy says that energy can change from one form to another, but it can never be created or destroyed.

I'm sure thankful for the chemical energy from this campfire.

The thermal energy from this hot chocolate warms my cold fingers.

I'd be thankful for some solar energy right now.

. . . or some radiant energy from the heater in my nice, warm bedroom at home.

Some Interesting Energy Facts

- The Sun is the most powerful source of energy in our solar system. It has provided heat and life for billions of years. Its energy is released by the burning of about 20 billion tons of hydrogen every minute.
- All living things depend on the Sun for their energy.
- All energy becomes heat. Rub your hands together and feel the heat that results from the energy you are using to move your hands together.
- Energy can change forms, but it cannot be lost.
- Energy can be transferred from one material to another. Food is warmed when heat energy from the oven is transferred into the food.
- Some energy is given off in every transfer. A light bulb uses energy to give light, but most of the energy it uses is given off as heat.
- Temperature is the measure of the average kinetic energy in the particles of an object.
- Specific heat is the amount of energy needed to raise the temperature of one kilogram of a material one degree Celsius. Different materials have different specific heats. The specific heat of water is 4190° C. The specific heat of copper is 380° C.
- Much of the energy used in the world comes from the burning of fuels such as coal, oil, natural gas, or gasoline.
- Energy is often used to generate electricity. The power of wind, or power from burning fuels is changed into heat and light energy.
- Energy is measured in joules (J). One joule is the amount of energy needed to move a 2-kilogram weight one meter in a second.

A joule is named after English scientist James Prescott Joule. Do you know why?

Yes! He was the one who first measured the transfer of thermal energy into an object.

The Obedient Can

You can make a can roll away from you and return when you call it!

1. Get a coffee can with a plastic lid. Use a can opener to punch a hole on each side of the end of the can.

2. Make one cut in a long, sturdy rubber band. Follow the diagram to thread it through the holes in the can.

3. Attach a heavy bolt to the center of the rubber band. Tie it with string.

4. Punch two holes in the plastic lid to match the holes you punched in the can. Thread the rubber band through those holes and tie them securely on the outside of the lid.

5. Roll the can away from you. When the can stops or slows down, command the can, "Come back!" The can will return to you.

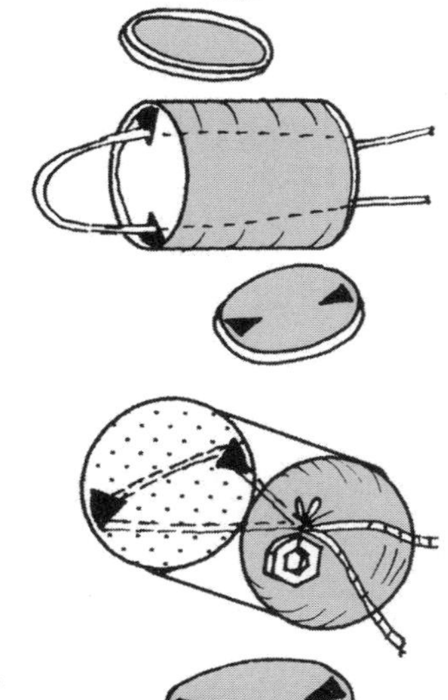

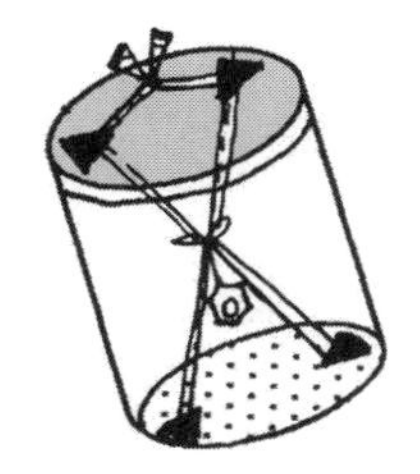

How It Works

When the can rolls away, the kinetic energy causes the rubber band to twist. The twisted band stores potential energy. When the can stops, it is ready to turn that potential energy back into energy of motion (kinetic energy).

Heat

Heat is energy that is transferred from an object at a higher temperature to one at a lower temperature. The energy of heat is called ***thermal energy***. Heat is measured in degrees. There are two commonly used scales: Celsius and Fahrenheit. On the Celsius scale, water freezes at 0° C and boils at 100° C. On the Fahrenheit Scale, water freezes at 32° F and boils at 212° F.

Transfer of Heat

Heat moves from one object or substance to another in one of three ways:

Conduction is transfer of heat from particle to particle when two substances come in contact with each other. Particles at high temperatures vibrate faster and further. They collide with the cooler particles and share their energy. Then the slower, cooler particles start to move faster.

When you put ice cubes into a drink, heat from the warmer drink transfers to the cubes.
When you put a pan of soup on a hot stove, heat from the stove transfers to the cooler soup.

A ***conductor*** is a substance that transfers heat. Some materials conduct heat better than others. Some good conductors are copper, iron, silver, and water. Plastic, gas, glass, and wood are poor conductors.

An ***insulator*** is a substance that reduces the transfer of heat. Poor conductors are often used as insulators.

Convection is transfer of heat by movement of matter. Convection is the way heat travels in fluids (liquids and gases). When a fluid is heated, its density and weight decreases. Cooler, denser fluid sinks, pushing the warmer, less dense fluid upwards. This motion of the fluid creates convection currents that spread heat throughout the substance.

Radiation is the transfer of heat that does not need matter. The heat from the Sun travels across millions of miles of empty space by radiation. Heat can transfer by radiation through air, as well. When you hold your hands out to warm them near a hot campfire (without touching the fire), you are enjoying heat transferred by radiation.

This cold is seeping through my clothes and chilling me to the bone!
That is not possible! Cold doesn't travel into a warmer place. Heat always moves from a warmer place to a cooler place. The heat is leaving you and seeping out through your clothes. That's why you are chilled.

Work & Machines

There is lots of work to be done in the world. It is a good thing we have machines, because much of the work is hard. Machines make all kinds of work easier.

Work is the transfer of energy as a result of motion.

A **machine** changes the amount or direction of the force that must be exerted to do work. A machine translates a small amount of force into a much larger force.

A **simple machine** is a machine that consists of only one part.

A **compound machine** is a combination of two or more simple machines. Many of the machines in the world combine several simple machines.

Effort force is the amount of force that must be applied to the machine.

Resistance force is the force exerted by the machine.

Mechanical advantage is the number of times a machine multiplies the effort force.

Power is the rate of doing work. It is the amount of work done per unit of time.

Work is measured in joules. To calculate work, multiply the force by the distance.

Power is measured in watts (W). To calculate power, divide the amount of work done by the amount of time it took to do it!

The Six Simple Machines

LEVER

The lever is a bar that is free to pivot about a fixed point called a fulcrum. A lever changes the amount and direction of the effort force.

PULLEY

A pulley is an arrangement of one or more wheels and rope. A pulley changes the amount of force and/or the direction of force needed to do work.

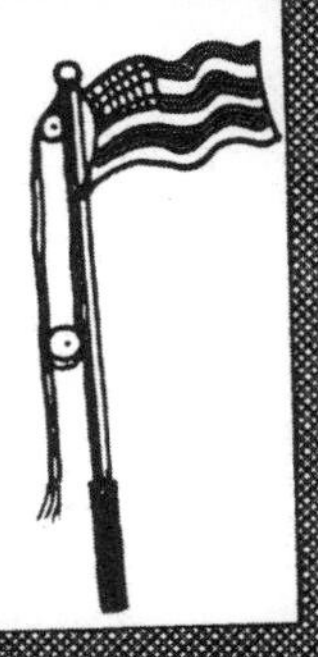

INCLINED PLANE

The inclined plane is a slanted surface used to raise objects. A pulley changes the amount of force needed to do work.

WHEEL & AXLE

This consists of a large wheel fixed to a smaller wheel called an axle. The two rotate together, changing the amount of force needed to do work.

SCREW

The screw is an inclined plane wound around a cylinder. It has a spiral thread along the edge. The screw changes the amount of force needed to do work.

WEDGE

The wedge is an inclined plane with at least one sloping side. The wedge changes the amount of force needed to do work.

Electricity

A phone rings. A flashlight beam helps you find your way to the tent. A siren warns of an approaching ambulance. A CD player booms out your favorite tune. The click of a mouse connects you to an awesome website. A flat screen on the grocery counter reads the price of your potato chips. A startling streak of light splits the dark sky during a thunderstorm. All of these are amazing tricks and displays of electricity.

Electricity is the energy resulting from the flow of electrons; and it's all around you every day, showing up naturally in spectacular displays and powering all sorts of conveniences.

An **electric charge** results when an object has too many or too few electrons. Every atom is made of particles that have electric charges. Electrons have a negative charge and protons have a positive charge. Electrons move through matter. When an object gains electrons it becomes negatively charged. An object that loses electrons becomes positively charged.

Static electricity is an electric charge built up in one place. Clothing and other objects can build up static electricity by rubbing against each other. You have seen this when you rub a ballon against clothing and place it near your head. Your hair stands on end!

A **conductor** is something that allows electricity to flow through it easily.

An **insulator** is a substance that is a poor conductor of electricity.

If life gives you lemons

Make Your Own Lemon Battery

You will need: light bulb, copper wire, 2 lemons, a knife, steel wire or paper clips

1. Squeeze and roll the lemons to get them juicy. Do this gently; do not break the skin.
2. Cut a 10-inch strip of each wire. Remove any insulation from the ends of the wires so they are bare. Cut a 3-inch strip of each wire. Twist these together.

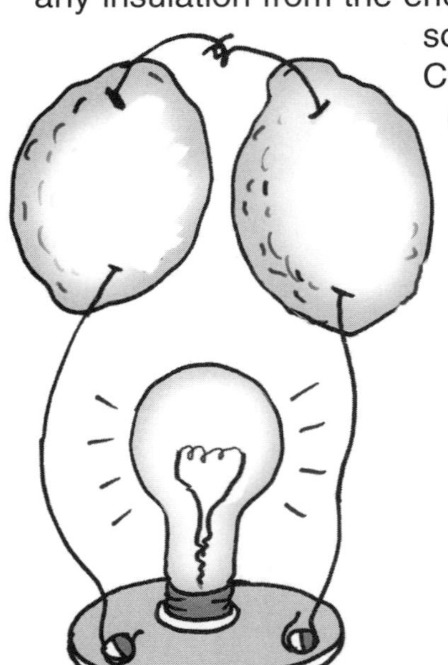

3. Attach one end of each wire to the light bulb holder. Place the other end of each wire deep into one of the lemons. Place the small copper-steel wire combination into the lemons as shown.

How It Works

The chemical reactions between the lemon juice and the two metals pushes electrons to flow through the circuit you have made. If the lemon battery is not strong enough to light the bulb, touch the two wires to your tongue to complete the circuit. You will feel the tingle of the electricity produced by the lemon battery.

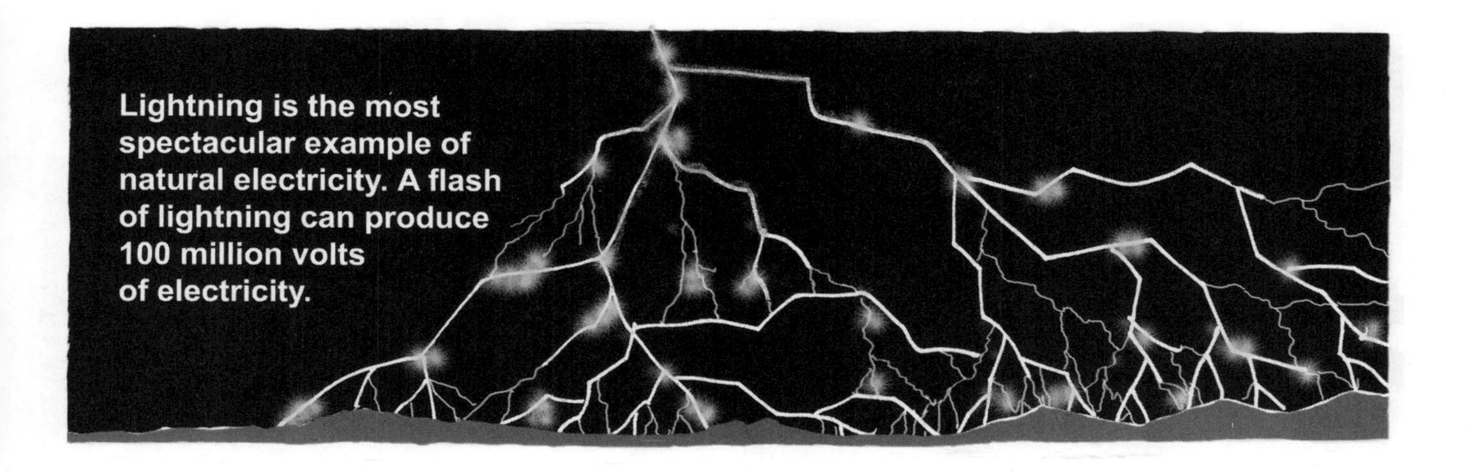

Lightning is the most spectacular example of natural electricity. A flash of lightning can produce 100 million volts of electricity.

Electric current is a steady flow of electrons or other charged particles through a conductor. For practical use, such as to run a machine or light a bulb, current needs to flow continuously. Some source of force, such as a battery, is needed to keep the electrons moving.

Direct current (DC) is the flow of electrons in one direction. When the force pushing the electrons is a battery, they flow from a negative terminal and flow to a positive terminal.

Alternating current (AC) is current that changes directions. In North America, power companies distribute alternating current that changes directions 120 times a second. This is called 120 current.

Potential difference is the difference in potential energy between the electrons in one place and the potential energy in the electrons in another place. *Voltage* is another word for potential difference.

Resistance is a measure of how hard it is to push electrons through a conductor. The resistance of a conductor depends on the thickness, length, and type of material of the conductor. Some materials, such as wood, glass, and plastic, have high resistance to the flow of electricity. Others, such as copper and water, have little resistance.

Electric power is the rate at which a device (such as a machine) changes electricity into another form of energy.

Electrical energy is the energy used by an electrical device. The amount of energy used by an electrical device depends on the amount of power delivered and the length of time the power is used.

Electric Conversations

What is a **coulomb?**

It's the charge on 6.24 billion, billion electrons. (That's 6.24 x 1,000,000,000 x 1,000,000,000.)

What is an **amp?**

An amp is short for ampere (A). It is the unit for measuring the rate of flow of electrical current. One ampere of current flows when one coulomb moves through a conductor in one second.

Tell me the significance of a **volt?**

A volt (V) stands for voltage. It is the measure of the force that pushes electricity through a conductor.

What is the unit for measuring resistance? It sounds something like humming!

Oh, um, that's an ohm!

What is a **watt?**

A watt (W) is the unit for measuring electrical power.
One watt equals one joule of work per hour.
A kilowatt (kW) is 1000 watts.
A hair dryer uses about 1500 watts of power.
A refrigerator uses about 700 watts.

I bet a 100-watt light bulb uses 100 watts at a time!

Benjamin Franklin
(1706-90)
was a powerful proponent and pioneer in the field of electricity. His experiment with a kite in a storm proved that lightning is an electrical phenomenon. Based on this knowledge, Ben created the first lightning conductor.

Watt did you say?

Electric Circuits

Get Sharp Tip #28

Electrons must be able to flow back to the battery or other source of force pushing them along.

I am <u>currently</u> studying to be an electrical engineer.

Do they <u>charge</u> a lot for that?

If it costs too much, I'll <u>switch</u> to another school.

A **circuit** is an unbroken path formed by electrical conductors. In order to flow, a current must have an uninterrupted loop of electrical conductors. The electricity will flow from negative to positive terminals in these batteries, if the wires are connected correctly. If a circuit is broken at any point, the electricity will stop flowing.

A ***series circuit*** has only one path for the current to follow. The current is the same in every part of the circuit. If any part of the circuit is broken, the current stops.

A ***parallel circuit*** has two or more separate branches for current to flow. If the circuit is broken in one branch, the current will still flow to other parts of the circuit.

A ***switch*** is a device on a circuit that can open and close the pathway to start and stop the flow of electricity.

series circuit

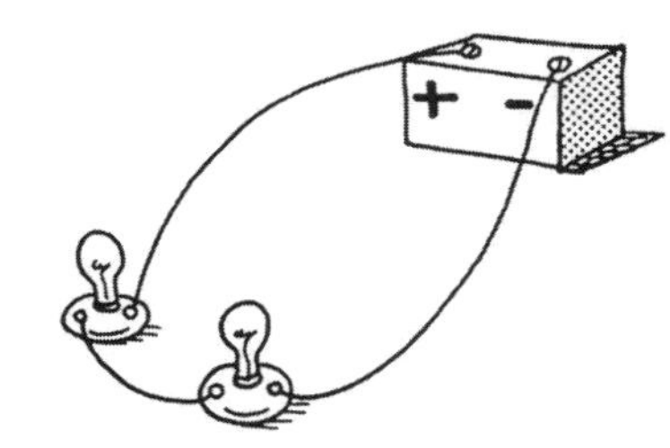

In this series circuit the lights will not burn as brightly because one battery is powering two bulbs.

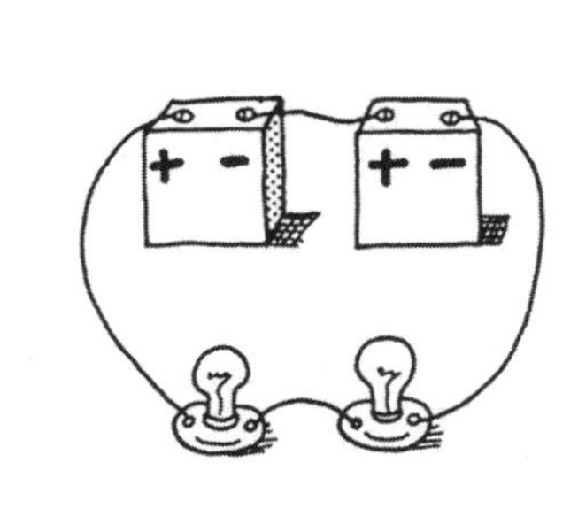

series circuit

The bulbs will not light in this series circuit because the wire is joining two negative terminals. A negative terminal must be joined to a positive.

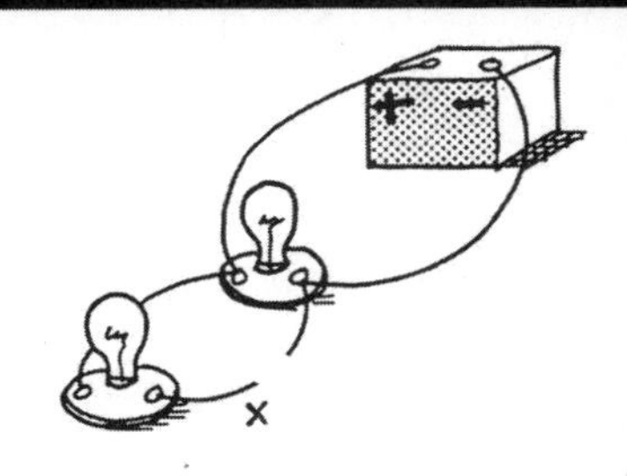

An open switch at point X in this parallel circuit would stop the flow of electricity through the second bulb, but not through the first.

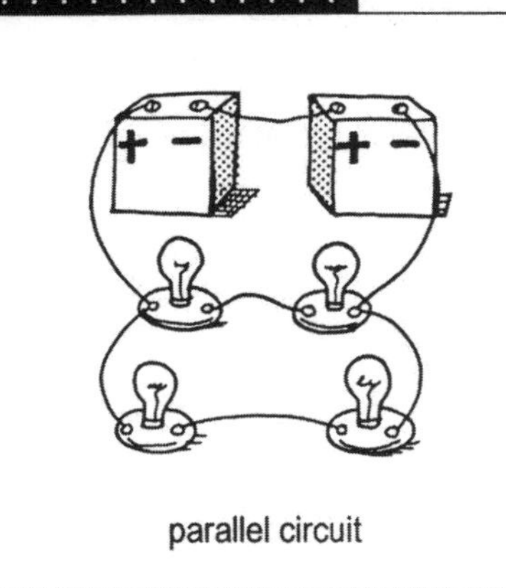

parallel circuit

Magnets

A ***magnet*** is an object that attracts other magnetic materials. The Earth is a giant natural magnet. Lodestone (magnetite), one of Earth's materials, is a natural magnet. Some other materials can be made into permanent magnets by exposing them to other magnets. These substances, such as iron, aluminum, cobalt, and nickel, can hold their magnetism for a long time. Also, some substances can be made into magnets by running electricity through them.

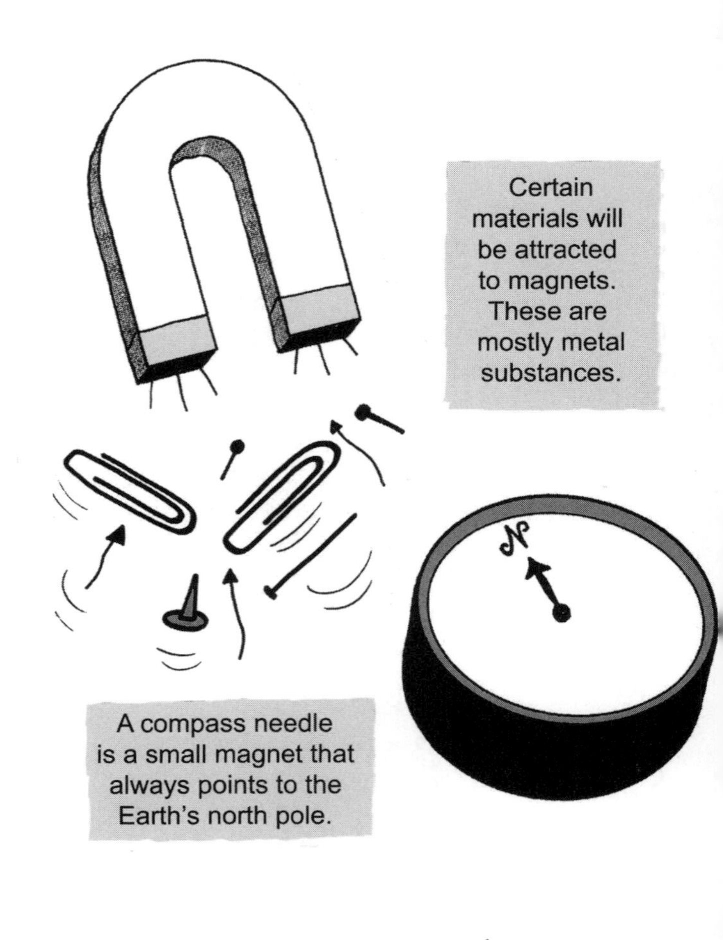

Certain materials will be attracted to magnets. These are mostly metal substances.
A compass needle is a small magnet that always points to the Earth's north pole.
N

Magnetic poles are the opposite ends of a magnet. The magnetic force is strongest at these locations. The poles are labeled *north* and *south*. When a magnet is held near a magnetic substance such as iron filings, most of the filings will cling to the magnet at its poles.

> ***Unlike poles*** (north and south) will attract each other.
>
> ***Like poles*** (north and north or south and south) will repel one another.

The **magnetic field** is a region around the magnet where the magnetic force acts. This region is near the poles, but extends out from the actual magnet.

N S

Like poles repel. Unlike poles attract.

S S

How to Make Your Own Magnetic Needles
Get a large needle. Hold it carefully in one hand. Stroke it against a magnet 20 times in one direction. Use only one end of the magnet; don't switch ends while stroking.
Now, use the needle to pick up some straight pins or paper clips.

Electromagnets

An ***electromagnet*** is a coil of wire with electric current flowing through it. Electromagnets have wide use in the world of electronics and other technology. They are used for heavy machinery and motors, and can be found working wonders in cars, trucks, trains, televisions, and tape recorders.

In an electromagnet, the ends of the coils are the ***poles***. Often an electromagnet has an iron ***core*** in the center of the coil. This makes the magnet stronger.

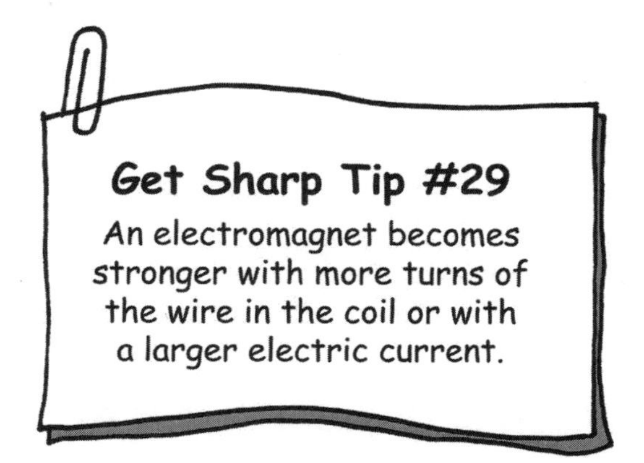

Get Sharp Tip #29

An electromagnet becomes stronger with more turns of the wire in the coil or with a larger electric current.

Electromagnetic Magic!

You will need:

a long nail · 2 D-size batteries · a small knife · string tape · pins & tacks · paperclips · 10 feet of insulated copper wire

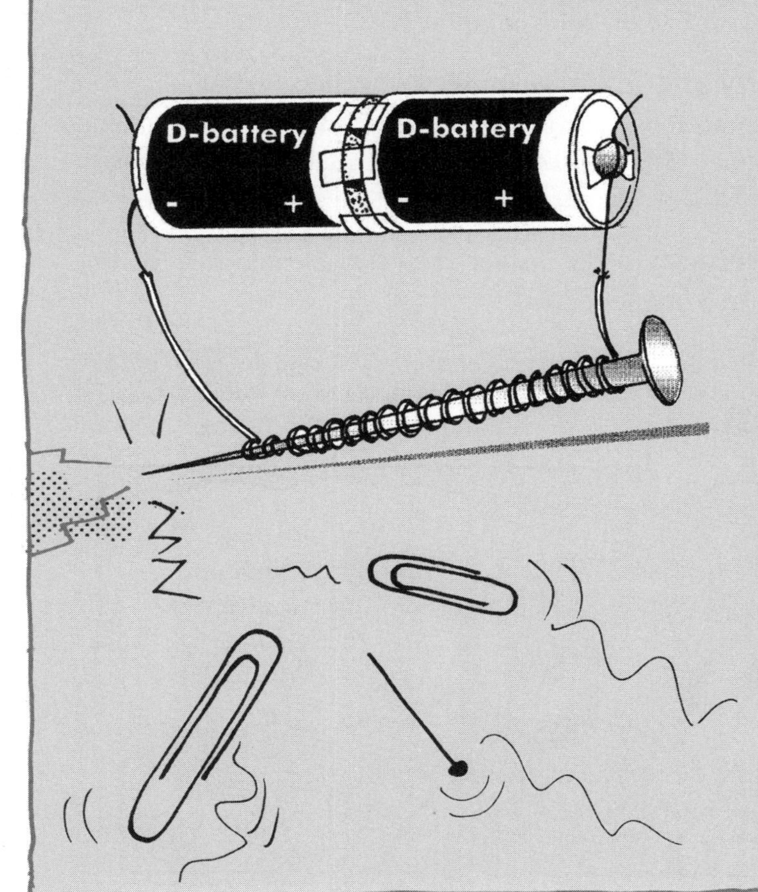

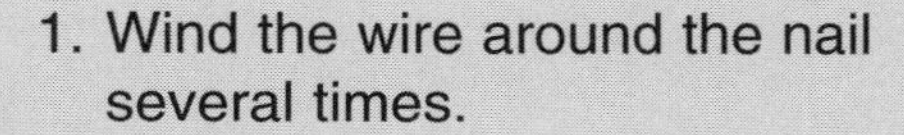

1. Wind the wire around the nail several times.
2. Leave 6 inches of wire sticking out at each end.
3. Use the knife to carefully cut the insulation off of the last inch at each end of the wire.
4. Tape the two batteries together end to end, with the bump on the bottom battery touching the bottom of the top battery.
5. Tape one wire end to the bottom of the battery. Tape the other wire to the top of the battery.
6. Your nail has become a magnet! Use it to pick up paper clips and pins. Try picking up other objects.

In 1820, Danish scientist Hans Christian Oersted discovered that electricity moving through wires could have a magnetic effect.

Waves

When you hear the word *waves*, you probably think of the rolling surf against a beach, whitecaps on a lake, or the lone surfer carried along on a giant wall of water. The waves that rise from water are only one kind of wave; there are many others. Light rays, radio waves, microwaves, X-rays, sound waves, and cosmic rays are a few of the other kinds of waves.

A ***wave*** is a rhythmic disturbance that carries energy. All waves transfer energy from one place to another. As it moves, the energy from a wave affects everything in the path of the wave.
In a ***transverse wave***, matter moves at right angles to the direction the wave is traveling.
In a ***compressional wave***, matter moves in the same direction as the wave travels.

Wave Characteristics

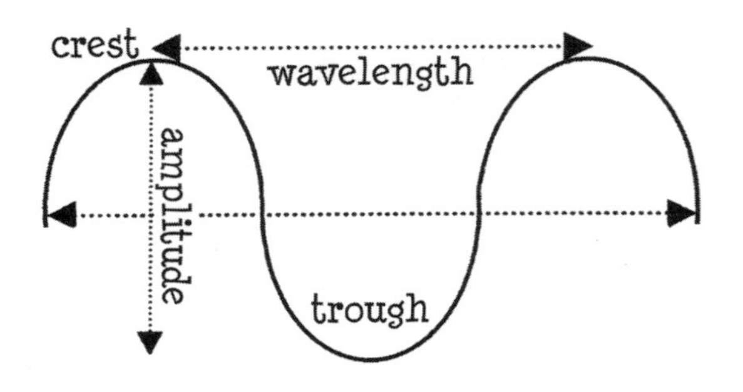

Waves with a long wavelength have low frequencies.

Waves with short wavelengths have high frequencies.

The **crest** is the highest point of the wave.

The **trough** is the lowest point of the wave.

The **wavelength** is the distance between one point on a wave the same point on the wave next to it.

The **amplitude** is the greatest distance particles in the wave rise or fall from their resting position.

Frequency is the number of waves that pass a given point in one second. Frequency is measured with a unit called a ***hertz*** (Hz). One hertz equals one wave passing a point per second.

Velocity is the distance traveled by any point on the wave in one second.

(Velocity equals wavelength multiplied by frequency.)

Vibration is the up and down movement of the wave.

What do scientists do at football games?
Go Atom U
They do the wave!

The Electromagnetic Spectrum

Electromagnetic waves are transverse waves that travel at the speed of light in a vacuum. As these waves travel, they transfer energy. The transfer of energy by electromagnetic waves is called ***radiation***.

The ***electromagnetic spectrum*** is an arrangement of the different waves according to their lengths. The only visible parts of the spectrum are light waves (rays).

Though radiation is not matter, the energy being transferred behaves like particles when it collides with matter. The "particle" of radiation is called a ***photon***.

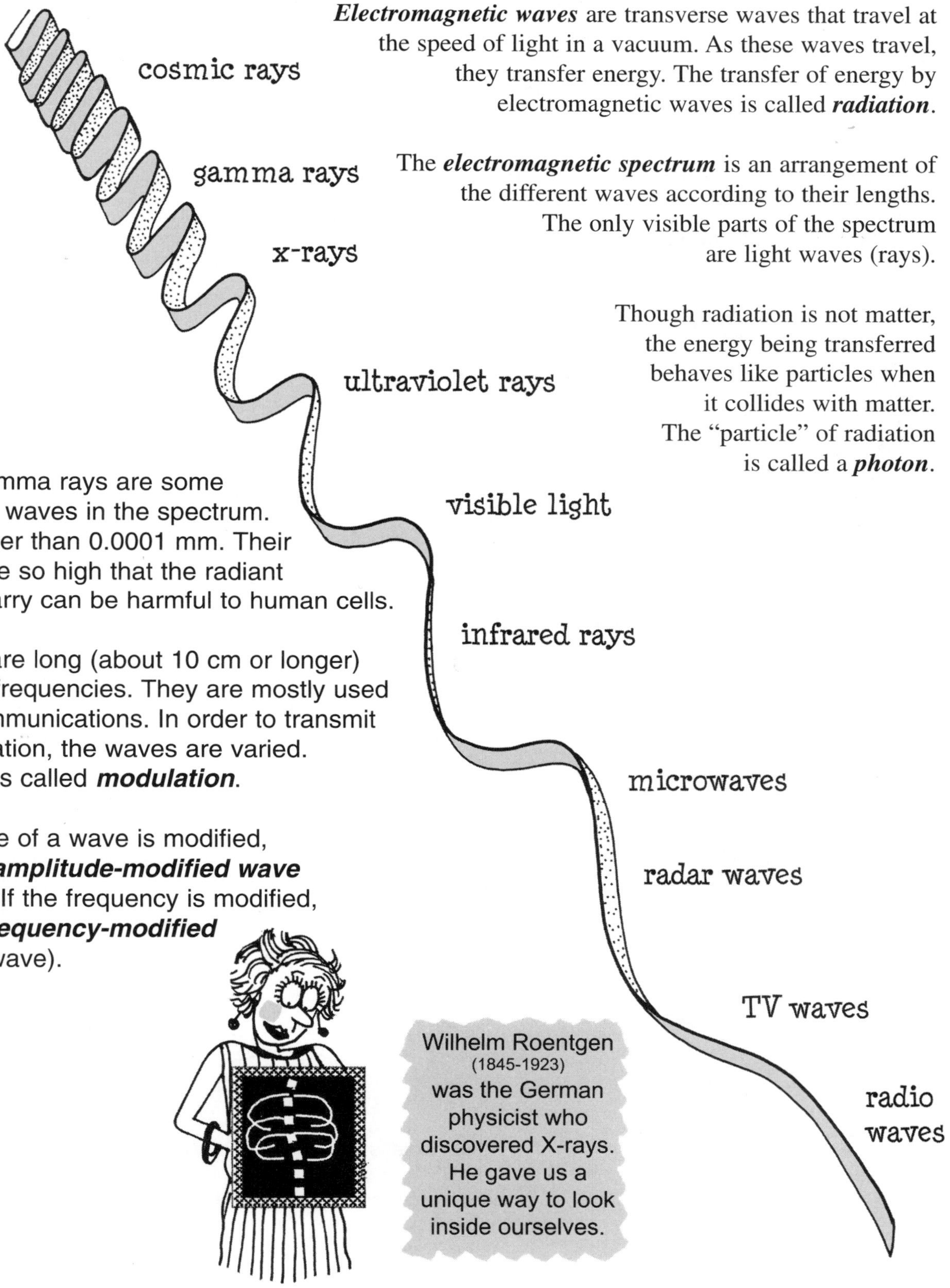

cosmic rays
gamma rays
x-rays
ultraviolet rays
visible light
infrared rays
microwaves
radar waves
TV waves
radio waves

X-rays and gamma rays are some of the shortest waves in the spectrum. They are shorter than 0.0001 mm. Their frequencies are so high that the radiant energy they carry can be harmful to human cells.

Radio waves are long (about 10 cm or longer) and have low frequencies. They are mostly used to transfer communications. In order to transmit a lot of information, the waves are varied. This variation is called ***modulation***.

If the amplitude of a wave is modified, it is called an ***amplitude-modified wave*** (or AM wave). If the frequency is modified, it is called a ***frequency-modified wave*** (or FM wave).

Wilhelm Roentgen (1845-1923) was the German physicist who discovered X-rays. He gave us a unique way to look inside ourselves.

Light

Light is the part of the electromagnetic spectrum that is visible. Our eyes can see a section of the spectrum containing a range of wavelengths that appear as different colors. White light is a mixture of all these colors. We see objects because they reflect light.

Reflection

When light waves are not absorbed by an object or do not pass through it, they may bounce off the object. This *bouncing off* action is called ***reflection***.

Light is reflected at an angle. The waves that strike the object are ***incident waves***. The angle at which they strike is the ***angle of incidence.***

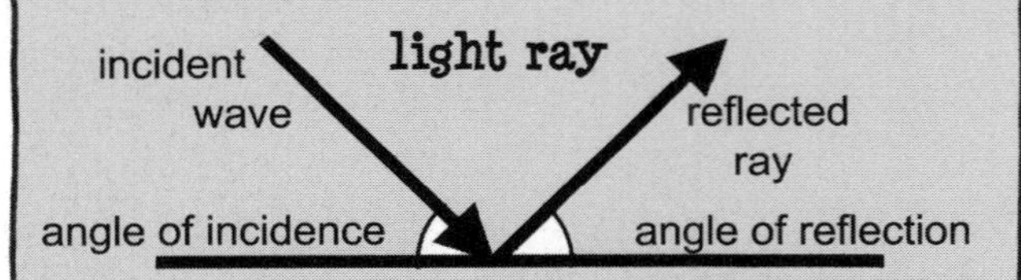

The waves that bounce off are ***reflected waves***. The angle at which they bounce off is the ***angle of reflection***. These two angles are always equal.

Refraction

Refraction is the bending of light waves. When light waves (rays) pass from one substance into another at any angle other than a right angle, they are bent. The bending distorts the image that is seen by the eye.

Lenses

A ***lens*** is a curved, transparent object. When light passes through a lens, the direction of the light is changed. Often the lenses produce an image that appears smaller or larger than the original object.

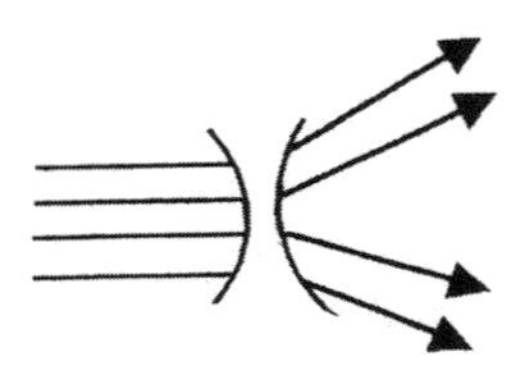

A concave lens is thinner in the middle. When light passes through it, the rays are bent away from each other. A concave lens can make objects appear smaller.

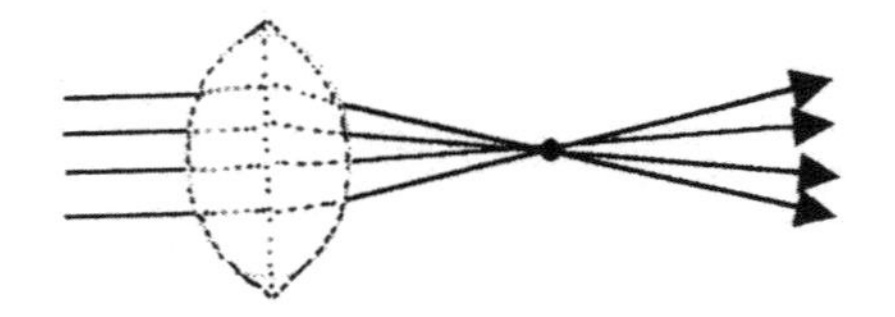

A convex lens is thicker in the middle. When light passes through it, the light is refracted, bent together to a focal point. A convex lens can be used to make objects appear larger.

Science Facts

When light hits an object, it passes through it, is absorbed by it, or bounces off of it.

A **transparent** object is made of material that allows the light to pass through it, and that you can see through it.

A **translucent** object is made of material that you cannot see through, but light can pass through it.

An **opaque** object absorbs light. Light does not pass through it and you cannot see through it.

How To Spy Around Corners

You will need:
- a quart-sized milk carton, - scissors,
- 2 identical pocket mirrors, - tape

1. Open the top of the milk carton.
2. Tape two mirrors at a 45° angle, parallel to each other, as shown.
3. The top mirror should be faced down. The bottom mirror should be faced up.
4. Cut a hole near the bottom, facing one mirror, as shown.
5. Cut a similar hole at the same location near the top, as shown.
6. Make sure that both mirrors and holes are in the same locations in relation to the top or bottom of the carton, as shown.
6. Use your periscope to see places without being seen yourself!

Why does it seem as if Sneaky Pete the Pirate can see around corners? **Because he can!** A periscope allows Pete to use the properties of light to see around corners. Follow the directions to make a periscope of your own.

I'll catch anyone who tries to steal my loot in the act!

Color

Different waves within the range of visible light have different lengths. These waves reflect with different colors.

A rainbow shows all the colors of the visible spectrum when drops of rain in the air refract sunlight. This is one time when we can see the red, orange, yellow, green, blue, indigo, and violet rays that combine to make white light.

Another way to see the colors of the light is to bend the light with a prism. We see individual colors in objects because the object has reflected the waves of that particular color.

Why is my skin green?
The molecules in the frog's skin absorb all the waves in light except the green ones. These rays are reflected back—and the frog looks green!
I'd rather be magenta!

Sound

The beat of a drum, a howling wind, a screeching owl, a whining child, a loud rock concert, a ringing phone, a sonic boom, a dripping faucet—these sounds and all others are made by the vibration of some object or objects. When something vibrates, it moves back and forth. The molecules in the object vibrate and cause the air around the object to vibrate, too. This sets up a sound wave that travels through the air in a series of ripples moving out from the object in all directions. Sound waves are ***compressional waves***, waves whose particles move in the same direction that the wave travels.

Sound cannot travel through a vacuum. This means that there is no sound in space. Sound must have matter for its transmission. The velocity of sound (the distance it travels in a given time) depends on the kind of matter through which it is traveling. Sound travels faster through liquids and solids than through gases.

The Doppler Effect

When a sound-making object or a listener is moving, there is a change in the frequency of the sound waves. This change in wave frequency is called the ***Doppler effect***. It affects the pitch of the sound (how high or low the sound is). The motion of a plane towards you crowds the sound waves together, increasing the frequency and causing the wavelengths to shorten. Shorter wavelengths cause the higher pitch you hear. When the jet moves past you, the waves are farther apart, the frequency of the waves is decreased, and the pitch appears lower.

The Doppler effect also occurs when the listener is moving. If you walk toward a lawn mower, you will find that the pitch becomes higher as you approach. The sound waves are striking your ears more frequently and wavelengths are shorter. As you pass the lawnmower, the pitch will be lower because the waves will strike your ears less frequently.

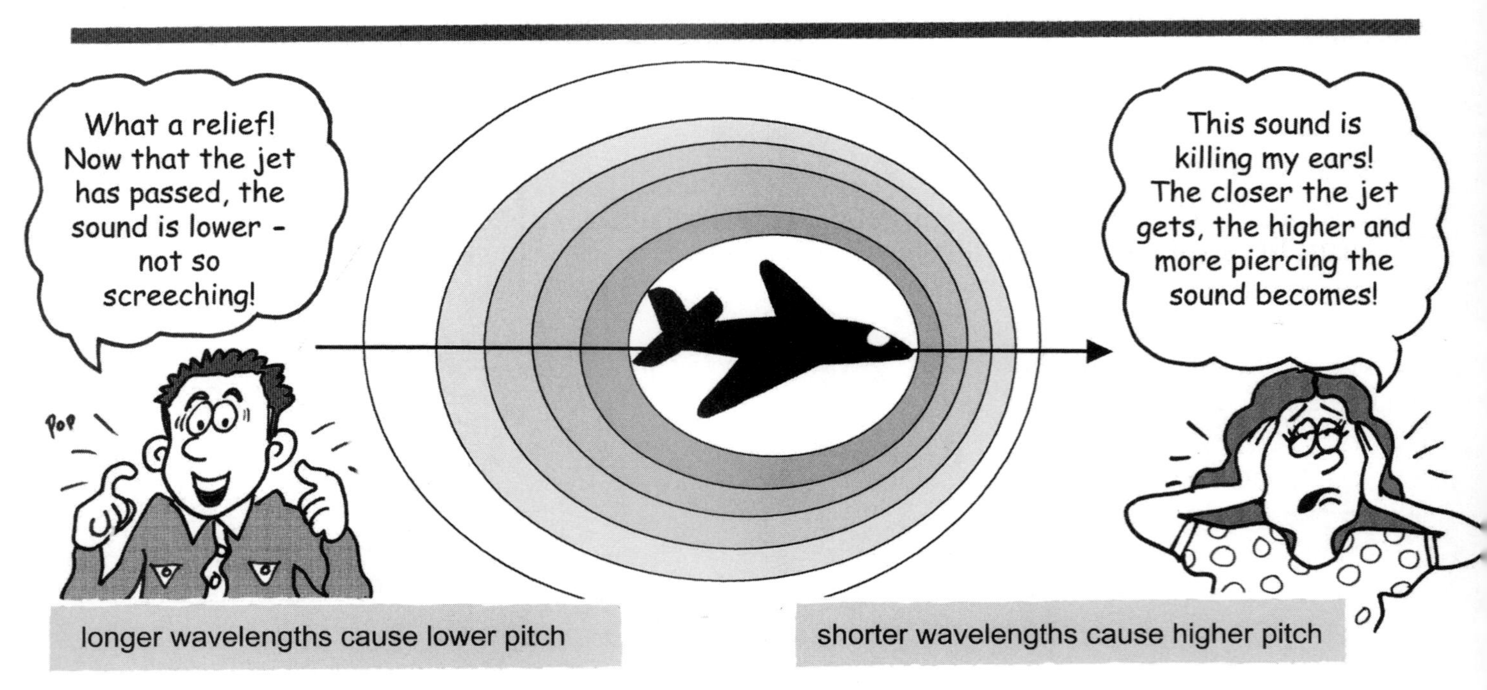

The Language of Sound

The **amplitude** of a sound is the greatest distance the particles in a wave rise or fall from their resting position. A sound wave with large amplitude will carry a **loud** sound.

The **intensity** (loudness) of a sound depends on the amplitude of the sound waves. The bigger the amplitude, the more intense the sound will be. The loudness of a sound can differ for different people. A sound that seems loud to one person may not be loud enough for someone else.

A **decibel (dB)** is the unit used to measure loudness or intensity of a sound. Normal talking has a loudness of about 30 dB. The quietest sound that can be heard is 0 dB. Sounds of 120 dB or above cause pain to the human ear.

The **frequency** of a sound is the number of waves that pass a certain point in a second. Frequency is measured in *hertzes* (Hz). Humans can hear sounds ranging from 20 to 20,000 Hz. Some animals can hear sounds with frequencies higher than those that can be heard by humans.

Pitch is the highness or lowness of a sound. The pitch of a sound is related to its frequency. Sounds with high pitches have high frequencies, and sounds with low pitches have low frequencies.

The **velocity** of a sound wave is the speed and direction it moves. Sound travels faster through liquids and solids than through gases. It travels at about 332 meters per second through dry air at 0° C. Sound travels faster through warm air than through cold air.

Tone quality has to do with the differences among sounds that have the same pitch and loudness. Tone qualities differ depending on the source of the sound. The tone quality of a note played on a piano is different from the tone of the same note played on a tuba.

Noise is sound that has no definite pitch or regular wave pattern. Noise can also be defined as unpleasant sound. Different people have different ideas about what sounds are pleasing.

Get Sharp Tip #30

Sounds above 20,000 Hz are called ultrasound.

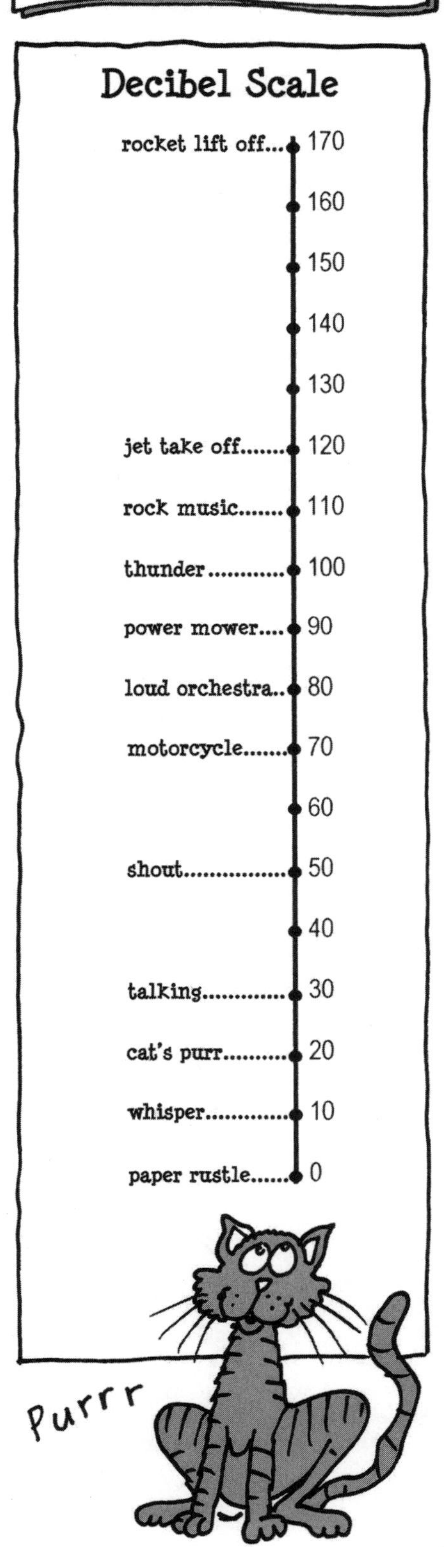

Fooling Around with Sound

Make your own stringed instrument and experiment with how different sounds are made.

You will need:

- a rectangular box (a cereal box or shoebox)
- rubber bands of different widths
- a thumbtack
- a wooden ruler, or a small block of wood
- a quarter

1. Stretch the rubber bands around the box. Line them up from thinnest to thickest.
2. Press the thumbtack into the lower edge of the box (as shown).
3. Use the quarter to pluck the strings of your instrument. Listen to the sounds. Pay attention to the pitch of the different sounds.
4. Place the wooden ruler (or block) under the rubber bands (as shown) to make a bridge. Listen to the sounds.
5. Wind one of the rubber bands around the thumbtack. This will tighten the tension on that band. Listen to the sounds made as you tighten the band more and more.

Find out:

which rubber bands produce the highest-pitched sounds

how the sounds change as the bands get shorter

how the sounds are different

how the sounds differ as the rubber band gets tighter

which bands produce low pitches

...Tra la la ...
....with a
banjo on my
knee.
plink
plunk

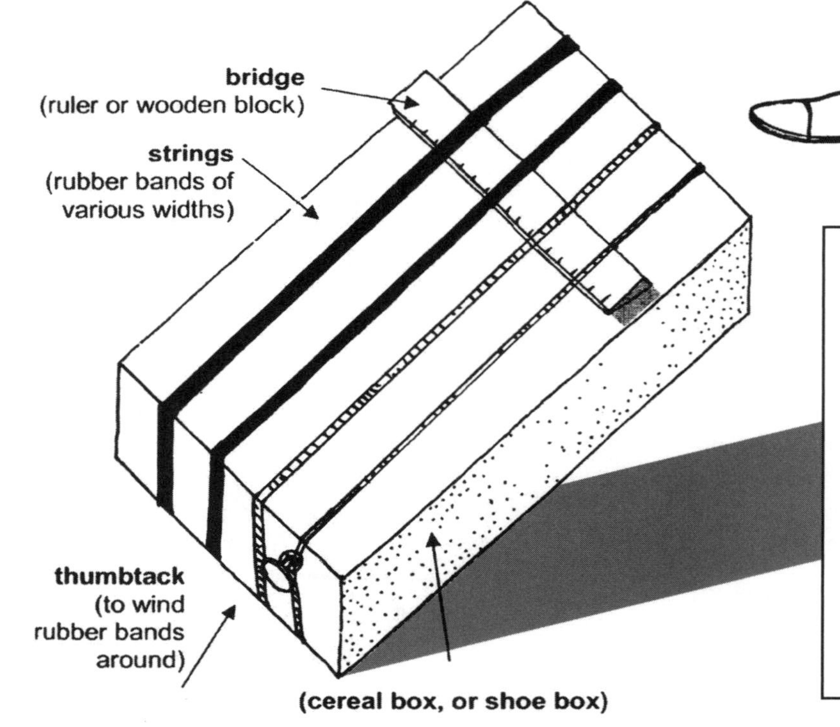

How it works:

As the rubber bands get thinner, shorter, or tighter, the pitch of the sound will be higher. This is because each of these makes the surface that is vibrating smaller, and results in faster vibrations. The faster the vibration, the higher the pitch. The wider, looser, longer rubber bands vibrate more slowly.

GET SHARP

on

SCIENCE TERMS

Get Sharp on Science Terms

A

abrading – scouring of rock surfaces by a glacier

abrasion – process in which wind carries fine particles that scour rock surfaces

absolute magnitude – the measure of a star's actual brightness

absorb – to take in or soak up

abyss – ocean depths of 2,000-6,000 meters

accelerate – to increase speed in movement

acceleration – the rate at which the velocity of an object changes

acid – a chemical substance that reacts with metals to release hydrogen; produces hydronium ions when dissolved in water

acid rain – pollution caused by waste compounds in the air forming weak acids in rain water

action-reaction pairs – a pair of forces which act on an object equally from opposite directions

active transport – movement of material by cells from areas of lower concentration to areas of higher concentration

active volcano – a volcano that erupts frequently

adaptation – a change in structure, function, or form that helps an organism adjust to its environment

adrenalin – hormone released by adrenal glands

aerobic exercise – exercise that strengthens the heart and lungs

aftershocks – tremors that follow an earthquake

air – the mixture of gases that makes up Earth's atmosphere

air mass – a body of air that has the same properties as the region over which it develops

air pressure – the pressure that air in the atmosphere exerts on everything on Earth's surface

algae – simple plants, without proper leaves, stems, or roots

alkali – a chemical substance that neutralizes acids

alternating current – current that changes direction

altitude – the height that something is above sea level

alveoli – tiny, thin-walled sacs in the lungs that are filled with air and surrounded by capillaries

ampere (amp) – the unit used to measure the flow of electrical current

amphibians – the class of vertebrates, including frogs, toads, and salamanders, that begins life in the water as tadpoles with gills and later develops lungs

amplitude – the greatest distance that the particles in a wave rise or fall from their resting position

anemometer – an instrument used to measure wind speed

angiosperm – a seed plant that produces seeds inside a fruit; a flowering plant

angle of incidence – angle at which incident waves strike an object

angle of reflection – angle at which reflected waves bounce off an object

antennae – a pair of movable, jointed sense organs on the heads of insects and other related organisms; used for taste, touch, and smell

antibodies – proteins made by the body that destroy poisons created by germs

antibiotic – substance produced by a living organism that slows down or stops the growth of bacteria

antiseptic – a substance that sterilizes and kills germs

aorta – largest artery in the body; carries blood away from the heart

aphelion – the point in an orbit where the body is farthest from the Sun

apparent magnitude – the brightness of a star as seen from Earth

apparent motion - the motion of an object relative to the position of its observer
The sun is sinking into the ocean.
The sun only seems to be sinking because you are standing on Earth. Actually, the sun is disappearing as Earth rotates away from it.

aquifers – permeable rocks containing water

arachnids – the class of arthropods, including spiders and scorpions, which have four pairs of legs, no antennae, and which breathe through lung-like sacs or breathing tubes

arthropoda – the phylum of invertebrate animals with jointed legs and a segmented body (such as insects, crustaceans, and arachnids)

arteries – vessels carrying blood away from the heart

asteroids – numerous small planets orbiting between the orbits of Mars and Jupiter

asthenosphere – the part of Earth's mantle that is most fluid

astronomy – the study of the stars, planets and other heavenly bodies

atrium – one of two upper chambers of the heart

atmosphere – the gaseous mass that surrounds any star or planet

atom – a tiny particle of matter consisting of a nucleus that contains protons and neutrons and an electron cloud that contains electrons

atomic mass – the number of protons plus the number of neutrons in the nucleus of an element

atomic number – the number of protons in the nucleus of an atom; identifies the kind of atom

auditory canal – tube leading from outer ear to inner ear

auditory nerve – nerve that carries sound vibrations and messages to the brain

aurora australias – streams of light in the upper atmosphere near the South Pole

aurora borealis – streams of light in the upper atmosphere near the North Pole

axis – imaginary line around which an object spins

B

balanced forces – two forces acting on an object with equal strength from opposite directions

barometer – instrument that measures air pressure

base – a substance that increases the hydroxide ion concentration when added to water

bedrock – the solid rock found under soil

biochemistry – the branch of chemistry that deals with plants and animals and their life processes

biome – an extensive community of plants and animals whose makeup is determined by soil, climate, and other geographical features

biosphere – regions of Earth's land, water, and atmosphere inhabited by living things

bladder – organ that stores urine

boiling point – the temperature at which a liquid turns into a gas

bone marrow – soft tissue at the center of bone; place where blood cells are made

botany – the study of plants

bronchi – tubes in the respiratory system that branch off of the trachea

bronchioles – smaller tubes into the lungs which branch off the bronchi

budding – a form of asexual reproduction in which a new organism grows from a bulge of tissue on the parent

buoyant force – the force with which a fluid pushes on objects

I believe that something other than a buoyant force is pushing up against me.
thump

C

cambium – plant tissue that makes stems grow thicker

camouflage – body coloring that protects an organism

capillary – one of many tiny (microscopic) blood vessels which join small arteries to veins

carbohydrate – any of certain nutrients made of sugar or starch

carbon dioxide – a colorless, odorless gas that is used by green plants and some protists in photosynthesis and which is given off by all living things in respiration

carbohydrates – energy-rich compound that comes from foods

cartilage – rubbery protein that cushions movable joints

cardiac muscle – muscle in the heart

catalyst – a substance that speeds up chemical reactions, but is not changed by the reaction

cause – anything that brings about a result

Celsius scale – a temperature scale used in the metric system at which water freezes at 0 degrees

central nervous system – brain and spinal cord

centrifugal force – a force acting on an object spinning around another to push the object outward

centripetal force – a force on an object operating toward the center of a circular path

cerebellum – part of the brain that controls balance and voluntary muscle action

cerebrum – largest part of the brain; controls thinking and awareness

change – the process of becoming different

I'm pretty sure I've produced a chemical change here.

chemical change - a change in which atoms and molecules form or break chemical bonds

chemical energy – energy released during a chemical reaction

chemical equation – a description of a chemical reaction using symbols and formulas

chemical formula – the combination of chemical symbols used as a shorthand for the name of a compound

chemical property – a property that describes the behavior of a substance when it reacts with other substances

chemical reaction – a change that produces one or more new substances

chemical symbol – the shorthand way of writing the name of an element

chemistry – the study of matter, its composition, and its changes

chlorophyll – the chemical in chloroplasts of plant cells that is needed for photosynthesis

chordata – the phylum of animals with an internal skeleton

chromosphere – the bright red layer of gas surrounding the Sun's photosphere

chromosomes – microscopic, rod-shaped bodies, which carry the genes that convey hereditary characteristics

chrysalis – pupa stage in insect development from the caterpillar stage

circuit – an unbroken path formed by electrical conductors

circulation – the movement of blood around the body due to pumping of the heart

classify – to put objects or processes in a group or category based on a common characteristic or group of characteristics

cleavage – breakage of a mineral along smooth, flat planes

climate – the average long-range weather of an area

cloud – tiny droplets of water grouped together in the atmosphere

cochlea – spiral-shaped ear structure; sound waves stimulate it to produce nerve impulses

cocoon – case of threads that an insect larva spins around itself

coefficient – a number that tells how many molecules of a substance are needed or produced in a reaction

coelenterata – the phylum of invertebrate animals with a central cavity and tentacles

colon – the last section of the large intestine

colloid – a heterogeneous mixture with particles of a size between that of a suspension and a solution

coma – the thick cloud of water and gases surrounding the nucleus of a comet

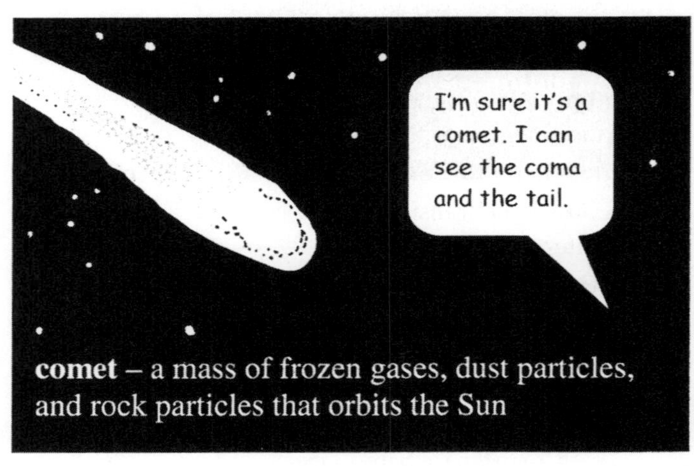

I'm sure it's a comet. I can see the coma and the tail.

comet – a mass of frozen gases, dust particles, and rock particles that orbits the Sun

commensalism – a relationship in which one organism lives on another without harming it

communicate – tell or show others the steps and results of an experiment

community – all the plants and animals that live together in a habitat

compound – a substance made up of two or more elements

compound machine – a machine made by combining two or more simple machines

compressional wave – a wave in which matter vibrates in the same direction that the wave moves

conductor – a material that transmits or carries electricity or heat

conduction – the transfer of heat from particle to particle when two substances come in contact

conglomerate – sedimentary rock made of pebbles and gravel cemented together by clay

conjugation – a method of reproduction that results in a zygospore

conifer – a seed plant that produces seeds in cones

conservation – all efforts to protect, replace, or make careful use of resources

conservation of energy – the principle that energy cannot be made or destroyed, but only changed in form, and that the total energy in a physical system cannot be increased or diminished

constancy – a state characterized by a lack of variation

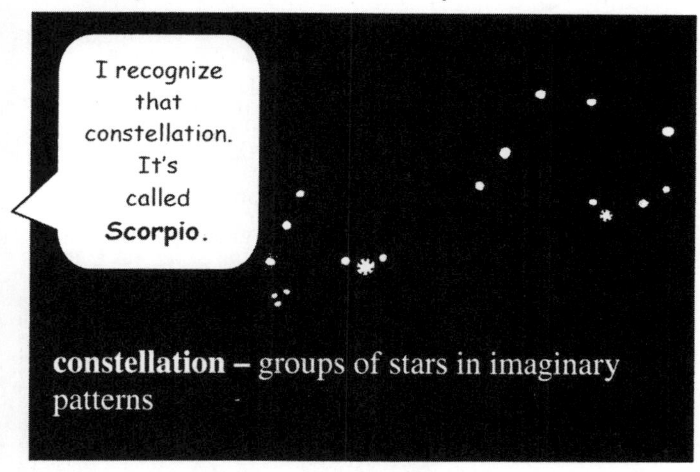
I recognize that constellation. It's called Scorpio.

constellation – groups of stars in imaginary patterns

convection – the transfer of heat by movement of matter

convection current – the movement of material within a fluid caused by uneven temperature; the upward movement of warm air and the downward movement of cool air

convergent boundaries – boundaries between plates in Earth's surface where plates collide

core – the center of the Sun; the outer atmosphere of the Sun

cornea – tough, protective outer covering of the eye

coulomb – the charge on 6.24 billion billion electrons

crater – basin-like depression at the top of a collapsed volcano

creep – the slow, downslope movement of a mass of Earth material

crust – the outermost layer of Earth, extending to a depth of about 35 km

crustaceans – the class of arthropods, including lobster, crabs, and shrimps, that usually live in the water, breathe through gills, and have a hard outer shell and jointed appendages

crystal – a solidified form of a substance in which the atoms or molecules are arranged in a definite pattern

cycle – a series of events or operations that occurs regularly and usually leads back to the starting point

cytology – the study of cells

D

deceleration – negative acceleration; the rate of change in velocity when the velocity decreases

decibel – the unit for measuring the loudness of sound

decomposer – an organism that helps another organism to decay

decomposition – weathering by chemical processes

deep water wave – a wave moving in water that is deeper than one-half its wavelength

deflation – process in which wind picks up loose material from the ground surface and moves it

delta – fan-shaped deposit of sediment at the mouth of a river

density – the ratio of the mass of an object to its volume

density currents – currents formed by the movement of more dense water toward an area of less dense water

dermis – second layer of the skin

diaphragm – muscle between chest cavity and abdomen; assists with breathing

dicotyledon – a flowering plant that has two leaves (cotyledons) inside the seed; leaves have a network of veins

diffusion – the movement of particles in solution from an area of greater concentration to an area of lower concentration

direct current – electric current that moves in one direction only

disintegration – weathering by physical processes

diurnal – an event occurring once a day; usually referring to tides

divergent boundaries – boundaries between plates in Earth's surface where two plates pull apart

dominant species – a species that gets more of the resources in an environment and therefore survives better

DNA – (deoxyribonucleic acid) the acid in chromosomes that carries genetic information

dormant volcano – volcano that is inactive but could become active again

Doppler effect – a change in the frequency of sound waves due to the movement of the listener or the object making the sound

dunes – piles of sand dropped when sand-carrying wind meets an obstacle

E

earthquake – vibrations in Earth's crust caused by the movement or breaking of rock in the surface

echinodermata – the phylum of marine invertebrate animals with a water vascular system and usually a hard, spiny skeleton and radial body (starfish, sea urchins, etc.)

eclipse – the passing of one object into the shadow of another object

ecology – the study of the relationship between plants, animals, and their environment

ecosystem – a system consisting of a community of animals, plants, and bacteria and their interrelated physical and chemical environment

effect – a result; an event or situation that follows from a cause

effort force – the amount of force that must be applied to a machine

electric charge – a positive or negative charge resulting when an object has too many or too few electrons

electric current – a steady flow of electrons through a conductor

electric energy – the energy used by an electrical device

electric power – the rate at which a device changes electricity into another form of energy

electricity – energy resulting from the flow of electrons

electromagnet – a coil of wire with current flowing through it that becomes a magnet

electromagnetic waves – transverse waves that travel at the speed of light in a vacuum

electromagnetic spectrum – an arrangement of the different electromagnetic waves according to their lengths

electron – a negatively charged atomic particle

element – a substance made up of only one kind of atom

elevation – the distance of a point above or below sea level

embryo – a young plant or fertilized cell

emulsion – a suspension of two liquids

endocrinology – the study of the endocrine system

energy-matter – the close relationships and interactions between matter and energy

evaporation – the change in state from solid or liquid to a gas

environment – the part of the biosphere surrounding a particular organism

epidermis – protective outer layer of skin

epiglottis – protective flap at the top of the trachea

equinox – a day when the hours of darkness and daylight are the same length

erosion – the moving of weathered particles on Earth's surface

Eustachian tubes – bony tubes that equalize pressure in the ear

exoskeleton – the hard outer covering protecting inner organs of an arthropod

experiment – a planned series of steps designed to find an answer to a question

external fertilization – the joining of a sperm and egg outside the body of the animal

extinct – volcano that is inactive for a very long time; form of life no longer living

F

Fahrenheit – the temperature scale in which the freezing point of water is 32 degrees and the boiling point of water is 212 degrees

fault – a fracture in a rock along which movement has taken place

fertilization – the joining of nuclei of the male and female reproductive cells

fission – splitting or breaking into parts

flagellum – the whip-like tail on some single-celled animals that helps in movement

floodplain – sediment dropped outside the river bed during flooding

fluid – a substance that flows, such as a liquid or gas

food chain – the path of food energy from one organism to another in an ecosystem

food web – a complete network of food chains

force – a push or a pull acting on something

form – the shape or other physical characteristics of an object, organism, or system

form & function – the relationship between the shape (form) of an object, organism, or system and its operation (function); usually a relationship where the function is dependent upon the form

fracture – break in rock or bones; irregular breakage of a mineral

fragmentation – a form of asexual reproduction where an animal divides into two or more pieces

frequency – the number of waves that pass a given point per second

freezing point – the temperature at which a substance changes from a liquid or gas to a solid

friction – a force that opposes motion between two surfaces that touch each other

front – a boundary where air masses meet

fulcrum – the point on which a lever is supported

function – the operation of an object, organism, or system

full Moon – Moon phase occurring when Earth is between the Sun and the Moon, with the Sun shining on the Moon so that it is visible from Earth

fungi – a kingdom of plant-like organisms that are parasites on living organisms or feed upon dead organic material, and which lack true roots, chlorophyll, stems, and leaves, and reproduce by means of spores

fusion – the joining together of parts into a whole

G

galaxy – a large grouping of millions of stars, planets, dust, and gases in outer space

gallbladder – organ that produces bile

galvanometer – a tool used for measuring very small electrical currents

gas – the form of matter that has no definite shape or volume

gemstones – a mineral or petrified substance that can be used as a gem when cut and polished

genes – units of inheritance passed from parents to offspring

genetics - the study of heredity
Genetics explains why you tend to look like you, whereas, I tend to look like me.

germination – a process where seeds begin to grow from an embryo into a seedling

geyser – a spring from which boiling water and steam shoot into the air at intervals

glacier – a moving river of ice and snow

glacial flow – the movement of glacial ice

gravity – the force of attraction that exists between all objects in the universe

groundwater – underground water

gymnosperms – a large class of seed plants which have the ovules borne on open scales (usually in cones) and which lack true vessels in the woody tissue (pines, spruces, cedars, etc.)

H

habitat – the type of environment suitable for an organism; native environment

hardness – a mineral's resistance to being scratched

hematology – the study of blood

heredity – the passing on of traits from parents to offspring by means of genes in the chromosomes

heterogeneous mixture – a mixture in which the composition is not the same throughout

hibernation – an animal behavior that involves a long period of rest or inactivity, usually in winter

homeostasis – the tendency of an organism toward balance

homogeneous mixture – a mixture in which particles of one substance are spread evenly throughout another substance

humidity – the ability of air to hold water

hydrosphere – all of the water on the face of the Earth

hypothesize – to make an assumption or a careful guess in order to test it further

I

incident waves – waves that strike an object

indicator – a device or substance that can be used to determine the acidity or alkalinity of a solution

inertia – the property of matter to resist changes in motion

infer – to draw a conclusion based on facts or information gained from an inquiry

igneous rocks – rocks formed from the cooling of hot, molten magma

inherited traits – traits that are passed on from parents to offspring

insoluble – that which cannot be dissolved

F
13
It's too bad that my report card is insoluble in water.

insulin – hormone that controls the amount of sugar in the bloodstream and the storage of sugar in the liver

intensity – the loudness of a sound

internal fertilization – joining of a sperm and egg that takes place inside the body of the female animal

interpret – to explain or tell the meaning of the results of an investigation

ion – an electrically charged atom that has lost or gained one or more electrons in a chemical reaction

iris – colored part of the eye; muscle that expands and contracts to let light into the eye

insulator – a substance that does not conduct heat or electricity well

investigation – an experiment or other organized plan for answering a question

J

jet stream – very rapid winds that move around the Earth from west to east at a high altitude

joints – place where a bone joins to another

joule – the unit used for measuring energy or work

K

kidneys – organs that remove waste from the blood

kinetic energy – energy of motion

kingdom – the major classification category of living organisms

L

landslides – quick movement of large amounts of material downhill

larva – the free-living, immature form of any animal that changes structurally when it becomes an adult; the second stage of insect development

larynx – voice box; part of the air passage between the mouth and nose and the trachea

lens – curved, transparent object that changes the direction of light; transparent disc in the eye that bends light to focus images

ligaments – strong, flexible fibers that hold bones together and stretch to allow bending at joints

liquid – the form of matter that has a definite volume but no definite shape

lithosphere – the outer rocky region of Earth that includes the crust and the rigid upper layer of the mantle

liver – organ that cleans wastes from the blood and stores useful substances

loess – a windblown deposit of fine dust particles gathered from deserts, dry riverbeds, or old glacial lakebeds

luminosity – the rate at which a star pours out energy

lunar eclipse – the partial or total obscuring of the Moon when Earth comes directly between the Sun and the Moon

M

machine – a device that changes the amount or direction of force that must be exerted to do work

magma – liquid or molten rock deep inside the Earth

magnet – a substance or object that attracts other magnetic objects to itself

magnetic field – a region around a magnet where magnetic force acts

magnetic poles – opposite ends of a magnet; place where magnetic force is strongest

main sequence period – the main life period of a star

mammal – a warm-blooded vertebrate that produces milk to feed its young; most develop inside the female's body before birth

mantle – the thick layer of the Earth between the crust and the core

marine – something that lives in the sea or is formed by the sea

mass – the amount of matter in an object

matter – anything that has mass and takes up space

measure – to compare an object or amount to a standard quantity in order to find out an amount or an extent

mechanical advantage – the number of times a machine multiplies the effort force

mechanical energy – potential and kinetic energy combined in lifting, stretching, or bending

medulla – the part of the brain at the base of the skull; controls involuntary muscle activities

meridian – imaginary lines running from Earth's north pole to south pole

metamorphic rocks – rocks that have been changed by heat and pressure

metabolism – the sum total of all the chemical processes and changes in an organism

metamorphosis – a process of going through structural changes to take the shape of an adult animal

meteor – the flash of light that occurs when a meteoroid is heated by its entry into the Earth's atmosphere; a shooting or falling star

meteoroid – any of the small, solid bodies that travel through outer space and are seen as meteors when they enter Earth's atmosphere

meteorite – the part of a meteoroid that falls to the Earth's surface

migration – an animal behavior that involves moving long distances to reproduce, mate, raise young, or find food

mineral – a naturally occurring, inorganic, crystalline solid with a definite chemical make-up

mitosis – the division of a cell nucleus

mixture – a substance containing two or more substances which are not in fixed proportions; the substances do not lose their individual characteristics when combined

molecule – the smallest particle of an element or compound that can exist in the free state and still retain the characteristics of the element or compound

model – a structure that visually represents real objects or events

mollusca – the phylum of invertebrates characterized by a soft, unsegmented body (often closed in a shell), and which usually has gills and a foot (oysters, snails, clams, etc.)

molting – a process by which an animal sheds its outer covering

Moment Magnitude Scale – a means for measuring the magnitude of earthquakes

momentum – the force produced by a moving body; the product of an object's mass and its velocity

monocotyledon – a flowering plant that has only one cotyledon (seed leaf) in its seed; has parallel veins

moraine – a huge mass of rocks, gravel, sand, and clay that has been deposited by a glacier

motion – a change in position of matter

mudflows – rapid movement of soil from weathering mixed with rain down a slope

mutualism — a dependent relationship in the environment that benefits both organisms

The bird eats the food scraps as he cleans the alligator's teeth. This relationship benefits both.

N

neap tides – low tides that occur when the Sun, Earth, and Moon form a right angle

nebulae – clouds of dust and gas where stars are born

negative charge – the charge of an atom having an excess of electrons

net force – a force resulting when the forces acting on an object are unbalanced

neuron – a nerve cell

neutral – neither positively nor negatively charged; neither acidic nor basic

neutron – a neutral atomic particle

new Moon – phase of the Moon in which the side of the Moon facing Earth is dark

non-electrolyte – a substance that will not make water conduct electricity

nonvascular plants – plants without vessels

nuclear energy – energy stored in the nucleus of atoms and released when atoms are split or fused

nucleus – the center of an atom which contains protons and neutrons; the solid part of a comet

nutrient – a chemical substance found in foods which is necessary for the growth or development of an organism

nutrition – the process of eating, digesting, and absorbing food; the study of healthy diets

nymph – the young stage in the development of insects that experience incomplete metamorphosis

O

observe – to recognize and note facts or occurrences; to watch carefully

ohm – the unit used for measuring resistance in an electrical conductor

orbit – the path of one object in free-fall around another object in space

order – the predictable behavior of objects, units of matter, events, organisms, or systems

offspring – a new organism produced by a living thing

optic nerve – nerve that carries messages from the eyes to the brain

orbit – the path a body follows in revolving around another body

organ – a group of specialized tissues that work together to perform a special function

organism – a living thing

organization – the arrangement of independent items, objects, organisms, units of matter or systems, joined into a whole system or structure

orogeny – the processes that build mountains

osmosis – diffusion of water through a membrane

ovaries – female organs that produce eggs

ovum – a female reproductive cell; an egg

oxidation – the union of a substance with oxygen; the process of increasing the positive capacity of an element or the negative capacity of an element to combine with another to form molecules; the process of removing electrons from atoms or ions

P

pancreas – organ that produces insulin which controls sugar in the blood

parallel circuit – a circuit with two or more separate paths for current to follow

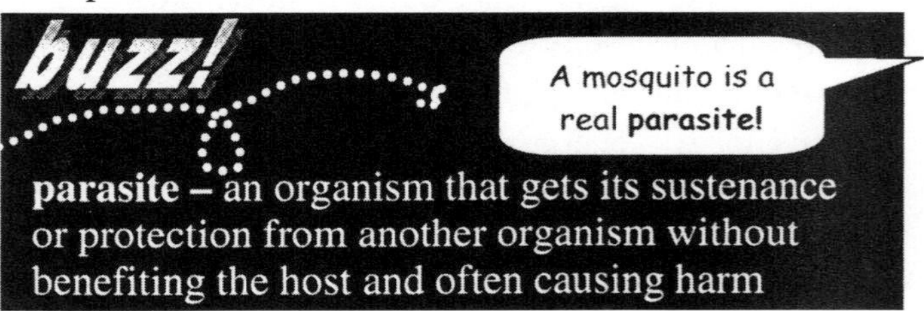

parasite – an organism that gets its sustenance or protection from another organism without benefiting the host and often causing harm

parathyroids – glands that regulate the balance of calcium in the bones and blood

partial eclipse – an eclipse in which the body (Moon or Sun) is only partially covered

penumbra – a partial shadow formed during an eclipse

perihelion – the point in an orbit where the body is closest to the Sun

period – a subdivision of a geologic era; the time taken for a cycle of events to take place; rows of elements in the periodic table

periodic table – an arrangement of elements in order of their atomic numbers

peripheral nervous system – nerves not including those in the brain and spinal cord

petal – the brightly-colored outer part of a flower

pH scale – a range of numbers used to tell the acidity or alkalinity of a solution

phases – any of the recurring stages of changes in the appearance of the Moon or a planet

phloem – plant tissues with tube-like cells that transport food from leaves to other plant parts

photon – a particle of radiation

photosynthesis – the process in which green plants use the Sun's energy to produce food

phylum – the largest classification category in a kingdom

physical change – a change in which chemical bonds are not formed or broken and no new substance is produced

physical property – a property that distinguishes one type of matter from another and can be observed without changing the identity of the substance

physics – the study of different forms of energy

pitch – a quality of sound (highness or lowness) determined by wave frequency

pituitary – master gland of the body; produces growth hormones

plain – a large, flat area with an elevation that differs little from that of the surrounding area

planet – an object in space that reflects light from a nearby star around which it revolves

plasma – fluid in which blood cells travel

plasmolysis – shrinking of cytoplasm in cells due to water loss

platelets – substances in the blood that produce clots

plates – huge sections of Earth's crust which lie beneath oceans and continents over a layer of molten rock in the mantle

plate tectonics – a theory that explains movements of continents and changes in Earth's crust caused by internal forces

plucking – the process of picking up fragments by a glacier and carrying them along

podiatry – the medical treatment of feet

pollen – the yellow, powder-like male reproductive cells formed in the anther of the stamen of a flower

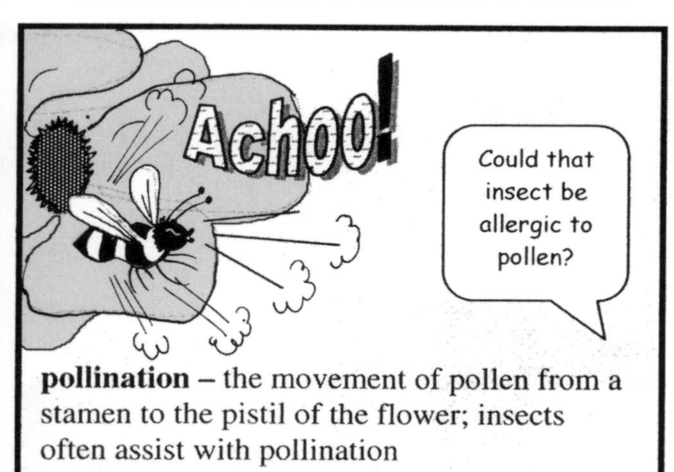

Achoo!
Could that insect be allergic to pollen?

pollination – the movement of pollen from a stamen to the pistil of the flower; insects often assist with pollination

population – the number of individuals of one species in a community

porifera – phylum of chambered animals (sponges) that live in water

positive charge – the charge of an atom having an excess of protons

potential difference – the difference in potential energy between the electrons in one place and the electrons in another

potential energy – energy due to position or condition

power – the rate of doing work

precipitate – an undissolved solid that usually sinks to the bottom of a mixture

precipitation – the falling of water or ice formed by condensation

predator – any animal that hunts and eats other animals

predict – to foretell what is likely to happen based on an observation or experiment

prehistoric – before recorded history

pressure – a pushing or squeezing force

prey – animals that are hunted and eaten as food

prominences – fingers of flame that surge from the Sun's chromosphere

property – a quality that describes or characterizes an object

proteins – organic compounds made of amino acids; important body nutrients necessary for life and growth in all organisms

protists – kingdom of simple organisms, mostly one-celled; most do not make their own food

proton – a positively charged particle found in the nucleus of an atom

protoplasm – the essential living material of all animal and plant cells

protozoa – the phylum of mostly microscopic animals made up of a single cell or group of identical cells and living mainly in water; many are parasites

pulmonary arteries – vessels that carry oxygen-rich blood from the lungs

pulmonary veins – vessels that carry carbon dioxide-laden blood to the lungs

pupil – tiny opening in the eye that lets light in

pulsar – rapidly rotating neutron star that gives out a beam of radiation which looks like a pulse

Q

quarantine – to separate an organism from others

quark – an subatomic particle

quasar – a quasi-stellar radio source; a starlike object that emits energy in waves

R

radiation – the transfer of energy by electromagnetic waves

receptors – nerve cells that receive impulses

reflection – bouncing off of waves or rays from an object

refraction – the bending of light as it passes from one medium to another

reproduction – the process by which organisms produce offspring

reptiles – cold-blooded animals with scales; live mostly on land and breathe air

resistance – any opposition that slows something down or prevents movement

resistance force – the force exerted by a machine

respiration – the process in which cells release energy from food

response – a change in an organism's behavior as a result of a stimulus

retina – screen of light-sensitive receptor cells which receive images in the back of the eye

revolve – to travel in an orbit around an object

Richter scale – a means for measuring the magnitude of earthquakes

riverbed – the path through which a river flows

river load – the material carried by a river

revolution – the movement of a body (or object) around another body (or object)

rotation – the turning or spinning of an object on an axis

runoff – water from precipitation that flows across Earth's surface and eventually returns to lakes, rivers, and oceans

S

salt – the substance that results when the right quantities of an acid and a base neutralize each other

satellite – a small planet that revolves around a larger one; a manmade object put into orbit around some heavenly body

saturated solution – a solution that holds all the solute it can at a given temperature

scavengers – animals that feed on dead organisms

scientific inquiry – a way of doing investigations and looking for explanations about happenings in the physical world; a series of steps generally followed in looking for answers to questions in science

seed – a ripe, fertilized ovule that will develop into a plant under suitable conditions

seedling – a young plant

sedimentary rocks – rocks formed by the cementing together of materials

semicircular canals – canals in the ear containing fluid; help keep balance

sense – a power that allows animals to be aware of their surroundings

seismic waves – vibrations set up by earthquakes

seismograph – an instrument that measures movements in the Earth's crust

sepals – the green, leaf-like structures that surround the bottom of flowers

series circuit – a circuit with only one path for current to follow

shallow water wave – a wave in water shallower than one-half its wavelength

shooting star – a briefly visible meteor

shoreline – the boundary where land meets the water of the ocean or lake

shore zone – area along the shore between the point of high tide and low tide

simple machine – a machine consisting of only one part

slump – a curved scar left when layers of rock slip downslope

smooth muscle – involuntary muscle present in walls of many internal organs

soil – a mixture of decayed organic material, weathered rock, air, and water, that covers much of the land on Earth

soil horizons – layers or regions of soil

solar eclipse – an eclipse that occurs when Earth is in the Moon's shadow

solar energy – energy that is trapped from the Sun

solar flares – sudden increases in brightness of the Sun's chromosphere

solid – a form of matter that has definite size and shape

solstice – a day when one of Earth's poles is tilted directly toward or away from the Sun

solubility – the amount of a substance that will dissolve in a specific amount of another substance at a given temperature

solute – the substance being dissolved in a solution

solution – a homogeneous mixture with tiny particles of a substance spread throughout another substance; particles cannot be filtered or settled out

solvent – the substance in which a solute is dissolved in solution

species – the smallest category in the kingdom in which only one kind of organism is classified

speed – the rate of motion of a body or object

sperm – male reproductive cell

spinal cord – thick cord of nerves that runs from the brain through the vertebrae

spring tide – a tide that occurs when the Sun, Moon, and Earth are aligned

spore – reproductive cell; grows into an organism

stamen – the male reproductive organ of an angiosperm

static electricity – electricity produced by charged bodies; charge built up in one place

stimulus – something happening in the environment that affects the behavior of an organism

stirrup, hammer, anvil – three bones of the inner ear that carry vibrations from sound to the auditory nerve

stratosphere – the second layer of the atmosphere (above the troposphere) which extends six to fifteen miles above the Earth's surface and where the temperature is fairly constant

subatomic particles – particles that make up and are smaller than atoms (protons, neutrons, electrons, quarks)

substance – a chemical element or compound with a known composition

sunspots – dark spots on the Sun where the temperature is cooler

supersaturated solution – a solution that holds more solute than normal at a given temperature

suspension – a cloudy mixture of two or more substances that settles upon standing

striated muscle – voluntary muscle made of bands called striations (also called skeletal muscle)

switch – a device on a circuit that can open and close the pathway for flowing current

symbol – the shorthand way to write the name of an element

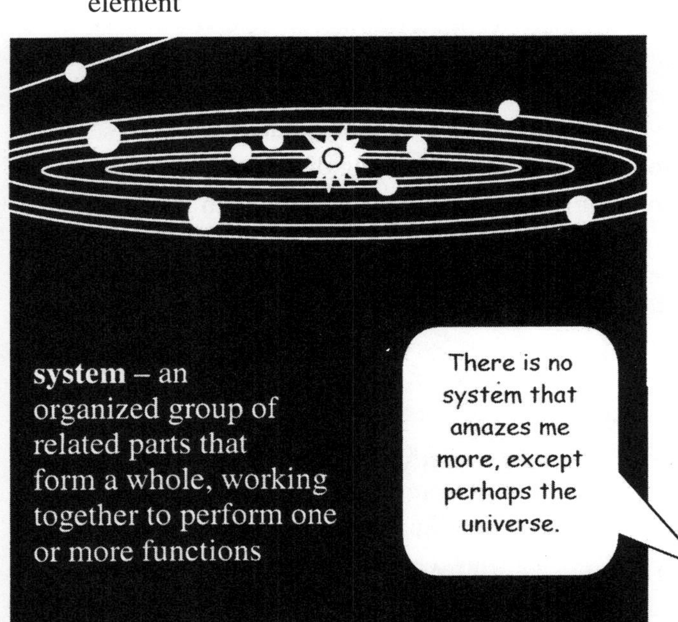

system – an organized group of related parts that form a whole, working together to perform one or more functions
There is no system that amazes me more, except perhaps the universe.

T

talus – eroded material that collects at the foot of a steep slope or cliff

taste buds – groups of receptor cells on the tongue sensitive to substances dissolved in saliva

telescope – an instrument which makes distant objects appear closer and larger

temperature – the measure of average kinetic energy of the particles in a material

tendons – strong bands of tissue that attach the ends of muscle fibers to bones

territoriality – actions designed to protect or defend a certain geographical area

testes – sperm-producing organ of the male reproductive system

thermal energy – the total energy in all the particles of an object; the amount depends on its temperature

thyroid – gland that controls the rate at which food is used in the body

tides – shallow water waves caused by the gravitational attraction among Earth, Moon, and Sun

till – unsorted, unlayered glacial deposit of boulders, sand, and clay

tissue – groups of similar cells that specialize to do a particular job in the body

tone quality – the difference among sounds that have the same pitch and loudness

topsoil – the highest, richest layer of soil (Horizon A)

total eclipse – an eclipse in which the Moon or Sun is entirely obscured from view for a period of time

trachea – windpipe; tube carrying air from the mouth to the lungs

trait – a characteristic

transform faults – boundaries where plates slide or scrape each other

translucent – material that transmits light but does not allow you to see through it clearly

transparent – having the property of transmitting or passing light

transpiration – the process through which organisms lose water

transverse wave – a wave in which matter moves at right angles to the direction the wave is traveling

tributaries – streams and small rivers that join into a large river

tropism – the response of a plant to a stimulus

troposphere – layer of atmosphere nearest Earth; contains about 90% of the gases in atmosphere

U

ultrasound – sound waves of a high frequency (above 20,000 Hz)

umbra – an inner, complete shadow formed during an eclipse

universe – the total of everything that exists in space

ureters – tubes that carry urine from the kidneys to the bladder

urethra – tube that carries urine out of the bladder and out of the body

uterus – female reproductive organ which holds a growing fetus

V

vacuum – the absence of matter

vapor – the gaseous form of a substance that is usually a solid or liquid at ordinary temperatures

vascular plants – plants with vessels

veins – vessels that carry blood toward the heart

velocity – the speed and direction of a moving object; the distance traveled by any point on a wave in one second

vent — tube-like passage through which magma and gases escape from a volcano

I guess even a volcano has to vent now and then.

ventricle – one of two lower heart chambers

vertebra – one of the small bones that make up the spinal column or backbone

vertebrates – animals with backbones

vibrations – rapid back and forth movements; the up and down movement of a wave

viscosity – the property of a liquid that describes how it pours

volt – the unit used for measuring the push of electricity through a conductor

W

waste – unwanted products of industrial processes, digestion, or respiration

water table – level in the ground below which water collects and cannot drain down any farther

watt – the unit for measuring electrical power

waves – ocean movements in which water rises and falls

weathering – the process by which surface rocks and other materials are broken down by wind, water and ice

wedge – a simple machine that is an inclined plane with either one or two sloping slides

weight – the force of gravity that a planet exerts on objects resting on the surface

wheel and axle – a simple machine that is a variation of a lever; consists of a large wheel fixed to a smaller wheel that rotate together

wind – movements of air parallel to the Earth's surface

work – the transfer of energy as a result of motion

wormhole – theoretical tunnels that link one part of space-time with another

X

X-ray – an invisible form of radiation with a very short wavelength; can pass through many materials that stop other rays, such as light rays

xylem – the tissue in plant roots and stems that supports the plant and carries water and nutrients to stems and leaves

Y

yeast – a type of fungus that feeds on sugary substances and produces by budding

Z

zone – one of the areas of Earth's surface that has a particular type of climate

zoology – the study of animals

zygote – a cell formed by fertilization

INDEX

D

E

O

P

Q

R